10th International Workshop on High-pT Physics in the RHIC/LHC Era 2014

AA006201

Journal of Physics: Conference Series Volume 805

Subatech Nantes, France
9 – 12 September 2014

Editors:

Gustavo Conesa Balbastre
Gergely Barnafoldi
Magali Estienne
Rachid Guernane
Andreas Morsch
Jan Rak
Alexandre Shabetai
Catherine Silvestre

ISBN: 978-1-5108-3755-3
ISSN: 1742-6588

Printed from e-media with permission by:

Curran Associates, Inc.
57 Morehouse Lane
Red Hook, NY 12571

Some format issues inherent in the e-media version may also appear in this print version.

Printed by Curran Associates, Inc. (2017)

For permission requests, please contact the Institute of Physics at the address below.

Institute of Physics
Dirac House, Temple Back
Bristol BS1 6BE UK

Phone: 44 1 17 929 7481
Fax: 44 1 17 920 0979

techtracking@iop.org

Additional copies of this publication are available from:

Curran Associates, Inc.
57 Morehouse Lane
Red Hook, NY 12571 USA
Phone: 845-758-0400
Fax: 845-758-2633
Email: curran@proceedings.com
Web: www.proceedings.com

Table of contents

Volume 805

10th International Workshop on High-p_T Physics in the RHIC/LHC Era
9–12 September 2014, SUBATECH Nantes, France

Accepted papers received: 31 January 2017
Published online: 7 March 2017

Preface

Heavy flavor and quarkonia

Photons

Initial conditions / saturation / event activity

QCD and medium effect

Jets

Jets, high pT hadrons and correlations

IOP Conf. Series: Journal of Physics: Conf. Series **805** (2017) 011001 doi:10.1088/1742-6596/805/1/011001

Preface

This volume contains selected papers presented at the *10th International Workshop on High-p_T Physics in the RHIC/LHC Era*. The workshop was held from 9-12 September 2014 at Subatech Nantes, France.

The first workshop of this series was organized 2006 in Trento, Italy. The purpose of the 10th workshop was to offer an opportunity for experimentalists and theoreticians to get together and discuss experimental results from RHIC and LHC - before the start of its secound run - and theoretical developments. Focus was also given to high-p_T physics in pA collisions especially at the LHC energies. Emphasis was given on discussion over a large scope : high-p_T physics, jets, photons, correlations, hard scattering and hard probes phenomena.

The main topics of the workshop were:

- Nuclear modifications of the parton distribution functions
- High p_T jet production in pp, pA and AA
- High p_T parton propagation in matter
- Nuclear modifications of the fragmentation functions
- Correlations with leading particles
- Direct photon and heavy flavor tagging
- Heavy flavors and Quarkonia
- Multi parton interactions

About 45 participants coming from 10 countries participated in the workshop. 8 invited talks and 21 oral contributions were presented at the conference.

IOP Conf. Series: Journal of Physics: Conf. Series **805** (2017) 011002 doi:10.1088/1742-6596/805/1/011002

Peer review statement

All papers published in this volume of *Journal of Physics: Conference Series* have been peer reviewed through processes administered by the proceedings Editors. Reviews were conducted by expert referees to the professional and scientific standards expected of a proceedings journal published by IOP Publishing.

Measurements of heavy-flavour decay leptons with ALICE

Denise Moreira de Godoy on behalf the ALICE Collaboration

SUBATECH - Ecole des Mines de Nantes, Université de Nantes, CNRS-IN2P3, Nantes, France

E-mail: moreira@subatech.in2p3.fr

Abstract. We present measurements of the nuclear modification factor of heavy-flavour decay leptons in p–Pb collisions at $\sqrt{s_{\rm NN}}$ = 5.02 TeV and Pb–Pb collisions at $\sqrt{s_{\rm NN}}$ = 2.76 TeV, the elliptic azimuthal anisotropy of heavy-flavour decay leptons in Pb–Pb collisions at $\sqrt{s_{\rm NN}}$ = 2.76 TeV, and the angular correlations between heavy-flavour decay electrons and charged particles in p–Pb collisions at $\sqrt{s_{\rm NN}}$ = 5.02 TeV and in pp collisions at $\sqrt{s}$ = 2.76 and 7 TeV. Results are compared to theoretical predictions.

1. Introduction

The ALICE experiment is dedicated to the study of the strongly-interacting medium formed in heavy-ion collisions at LHC energies. Heavy quarks, i.e. charm and beauty, are suited to access the properties of the created medium, since they are mainly produced in initial hard parton scattering processes and experience the whole evolution of the system.

The modification of the transverse momentum spectra of heavy-flavour hadrons in Pb–Pb collisions with respect to binary scaled pp collisions is sensitive to the parton energy loss within the Quark-Gluon Plasma (QGP). The measurement in p–Pb collisions is aimed at accessing cold nuclear matter (CNM) effects on heavy-flavour production, such as nuclear modification of the PDF at low x. The nuclear modification factors in Pb–Pb collisions ($R_{\rm AA}$) and in p–Pb collisions ($R_{\rm pPb}$) are defined as:

$$R_{\rm AA}(p_{\rm T}) = \frac{1}{\langle T_{\rm AA}\rangle}\frac{{\rm d}N_{\rm AA}/{\rm d}p_{\rm T}}{{\rm d}\sigma_{\rm pp}/{\rm d}p_{\rm T}},\quad R_{\rm pPb}(p_{\rm T}) = \frac{1}{A}\frac{{\rm d}\sigma_{\rm pPb}/{\rm d}p_{\rm T}}{{\rm d}\sigma_{\rm pp}/{\rm d}p_{\rm T}}, \tag{1}$$

where $dN_{\rm AA}/dp_{\rm T}$ is the $p_{\rm T}$-differential yield in Pb–Pb collisions, $T_{\rm AA}$ is the nuclear overlap function in Pb–Pb collisions, which is evaluated with the Glauber model [1], A is the mass number of the Pb nucleus, $d\sigma_{\rm pp}/dp_{\rm T}$ and $d\sigma_{\rm pPb/dp_T}$ are the $p_{\rm T}$-differential cross sections in pp and p–Pb collisions, respectively. For particles produced in hard-scattering processes, one expects nuclear modification factors equal to unity in absence of initial and final state effects in nuclear collisions.

The anisotropy of particle emission can be used to further investigate the QGP properties. The anisotropy is obtained from the particle azimuthal distribution in momentum space expressed as a Fourier series with respect to the reaction plane (RP), which is defined by the impact parameter direction and the beam direction [2]:

$$E\frac{{\rm d}^3N}{{\rm d}^3p} = \frac{1}{2\pi}\frac{{\rm d}^2N}{p_{\rm T}{\rm d}p_{\rm T}{\rm d}y}\left[1+2\sum_{n=1}^{\infty} v_n\cos[n(\varphi-\Psi_{RP})]\right], \tag{2}$$

where φ is the azimuthal angle of the particle emission direction, $\Psi_{\rm RP}$ is the (true) reaction-plane angle, which is experimentally estimated from the event-plane angle. The coefficients v_n, which can also be computed as $v_n = \langle\cos[n(\varphi - \Psi_{RP})]\rangle$, represent the azimuthal anisotropy magnitude of the n-th harmonic. The elliptic azimuthal anisotropy (v_2) of heavy-flavour hadrons at low transverse momentum provides insight into the collective motion of heavy quarks, whereas v_2 of heavy-flavour hadrons at high transverse momentum is sensitive to the path-length dependence of the heavy-quark energy loss within the QGP.

Another observable used to access the collective patterns in the particle production is the two-particle angular correlation, which is the differential yield of trigger-associated particle pairs in azimuth ($\Delta\varphi$) and in pseudorapidity ($\Delta\eta$):

$$C(\Delta\varphi, \Delta\eta) = \frac{1}{N_{\text{trigger}}} \frac{\mathrm{d}^2 N_{\text{pair}}}{\mathrm{d}\Delta\varphi \mathrm{d}\Delta\eta}, \tag{3}$$

where N_{trigger} is the number of trigger particles and N_{pair} is the number of pairs of trigger and associated particles. The angular correlation between heavy-flavour decay electrons (trigger particle) and charged particles (associated particle) in pp collisions can be used to estimate the relative contribution of beauty-hadron decay electrons to the yield of heavy-flavour decay electrons. The measurement in p–Pb collisions can be used to investigate cold nuclear matter effects and a potential existence of collective phenomena in such collisions.

In this paper, heavy-quark production is investigated by measuring electrons and muons from semi-leptonic decay channels, which have a branching ratio of the order of 10%. We present measurements of the nuclear modification factor of leptons from heavy-flavour hadron decays in p–Pb collisions at $\sqrt{s_{\text{NN}}}$ = 5.02 TeV and Pb–Pb collisions at $\sqrt{s_{\text{NN}}}$ = 2.76 TeV, the elliptic azimuthal anisotropy of heavy-flavour decay leptons in Pb–Pb collisions at $\sqrt{s_{\text{NN}}}$ = 2.76 TeV, and the angular correlations between heavy-flavour decay electrons and charged particles in pp collisions at $\sqrt{s}$ = 2.76 and 7 TeV and in p–Pb collisions at $\sqrt{s_{\text{NN}}}$ = 5.02 TeV.

2. ALICE experiment and particle identification

The ALICE detector is described in detail in [3]. The centrality [4] and the event-plane angle in Pb–Pb collisions are measured with the V0 detector, which comprises two scintillator counters positioned on each side of the collision vertex: V0A ($2.8 < \eta < 5.1$) and V0C ($-3.7 < \eta < -1.7$). The amplitude of the signal in the V0A detector is also used in p–Pb collisions to classify events according to the multiplicity of particles produced at backward rapidity.

Electrons are identified at mid rapidity ($|\eta| < 0.9$) with the Inner Tracking System (ITS), the Time-Projection Chamber (TPC), the Time of Flight (TOF), and the ElectroMagnetic Calorimeter (EMCal). The identification with the ITS and the TPC is based on the specific particle energy loss $\mathrm{d}E/\mathrm{d}x$. The time of flight of particles is used to reject hadrons in the low momentum region where the hadron $\mathrm{d}E/\mathrm{d}x$ curves cross the curve corresponding to electrons measured with the TPC. The electron identification with the EMCal is based on the distribution of the energy-to-momentum E/p ratio (with $E/p \approx 1$ for electrons) and the shower shape. The electron candidate sample is composed of heavy-flavour decay electrons and background electrons, which are mainly electrons from photon conversions in detector material and from Dalitz decays. The yield of background electrons is estimated using two different methods, depending on the analysis, and subtracted statistically from the identified electron sample. The first method estimates background electrons from invariant mass of electron-positron pairs, which is expected to have small values for pairs from conversions and from Dalitz decays. The second method uses a Monte Carlo (MC) calculation of the background contributions based on the measured π^0 and η-meson cross sections [5]. The selection of electrons from beauty-hadron decays is based on the track impact parameter (d_0), which is the distance of closest approach (DCA) between the charged track and the interaction vertex in the transverse plane. Due to the long lifetime of beauty hadrons ($c\tau \approx 500\ \mu$m), the impact parameter of beauty-hadron decay electrons is on average larger than the one of background electrons, which also include electrons from charm-hadron decays in this case.

Muons are reconstructed with the Muon Spectrometer at forward rapidity ($-4 < \eta < -2.5$). The track candidate in the muon tracking chambers is required to be matched with the trigger chambers in order to

IOP Conf. Series: Journal of Physics: Conf. Series **805** (2017) 012001 doi:10.1088/1742-6596/805/1/012001

reject the hadronic background. A cut on $p \times$ DCA, where DCA is the distance between the extrapolated muon track and the interaction vertex, is applied to remove non-primary tracks. The muon candidate sample includes muons from heavy-flavour hadron decays and muons from background sources, which are mainly π and K decays. Background muons are estimated in pp collisions via MC simulations, while the estimation in p–Pb and Pb–Pb are obtained with the measured muons from π and K decays at mid rapidity extrapolated to forward rapidity based on MC simulations. The forward-backward asymmetry of charged particles measured by the CMS Collaboration [6] is used for the extrapolation in p–Pb analysis.

3. Results

For all results shown in this section, the vertical error bars represent the statistical uncertainties, the horizontal error bars indicate the bin widths and the boxes represent the total systematic uncertainties. The boxes centered at $R_{\rm pPb} = 1$ and $R_{\rm AA} = 1$ represent the normalization uncertainty. The convention used for the rapidity sign in p–Pb collisions is negative for Pb-going direction, which addresses the orientation of the muon spectrometer, and positive for p-going one.

3.1. Nuclear modification factors

The pp reference for the $R_{\rm AA}$ and $R_{\rm pPb}$ measurements (Eq. 1) of electrons from heavy-flavour hadron decays at $\sqrt{s_{\rm NN}}$ = 2.76 and 5.02 TeV, respectively, was obtained by using a pQCD-driven $\sqrt{s}$-scaling [7] of the measured cross section of heavy-flavour decay electrons in pp collisions at $\sqrt{s}$ = 7 TeV [5] for $p_{\rm T} < 8$ GeV/c and by using FONLL predictions [8] for $p_{\rm T} > 8$ GeV/c. The measured cross section of electrons from beauty-hadron decays in pp collisions at $\sqrt{s}$ = 7 TeV [9] was scaled to 5.02 TeV and used as reference for $R_{\rm pPb}$ of electrons from beauty-hadron decays at $\sqrt{s_{\rm NN}}$ = 5.02 TeV. The pp reference for the $R_{\rm pPb}$ of heavy-flavour decay muons was obtained by $\sqrt{s}$-scaling [7] of the measured cross section of muons from heavy-flavour hadron decays in pp collisions at $\sqrt{s}$ = 2.76 [10] and 7 TeV [11]. The measured cross section of heavy-flavour decay muons in pp collisions at $\sqrt{s}$ = 2.76 [10] was used as reference for $R_{\rm AA}$ of heavy-flavour decay muons at $\sqrt{s_{\rm NN}}$ = 2.76 TeV.

Figure 1 (left panel) shows the nuclear modification factors measured for electrons from heavy-flavour hadron decays and from beauty-hadron decays in minimum bias p–Pb collisions at $\sqrt{s_{\rm NN}}$ = 5.02 TeV. Results are compatible with unity within the uncertainties. The right panel of Fig. 1 shows the comparison of the measured $R_{\rm pPb}$ of electrons from heavy-flavour hadron decays to FONLL calculations [8] with EPS09 parameterization [12] for the nuclear modification of the PDFs in p–Pb collisions. The calculation describes the data within the uncertainties.

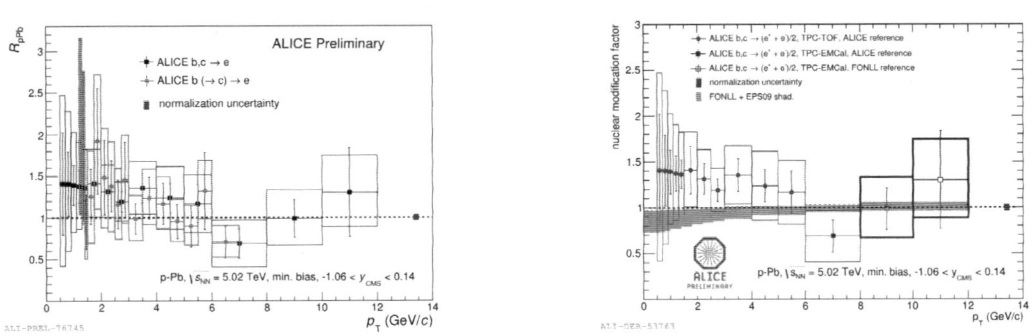

Figure 1. Left panel: $R_{\rm pPb}$ of electrons from decays of heavy-flavour hadrons (square symbols) and beauty hadrons (circle symbols) as a function of $p_{\rm T}$ at $\sqrt{s_{\rm NN}}$ = 5.02 TeV at mid rapidity ($-1.06 < y_{CMS} < 0.14$). Right panel: $R_{\rm pPb}$ of electrons from heavy-flavour hadron decays as a function of $p_{\rm T}$ at $\sqrt{s_{\rm NN}}$ = 5.02 TeV compared to FONLL calculations [8] with EPS09 parameterization [12] for the nuclear modification of the PDFs in p–Pb collisions.

Figure 2 shows the nuclear modification factors measured for muons from heavy-flavour hadron decays in minimum bias p–Pb collisions at $\sqrt{s_{NN}}$ = 5.02 TeV at forward ($2.5 < y_{CMS} < 3.54$) and backward ($-4 < y_{CMS} < -2.96$) rapidities. Results are also compared to NLO(MNR) pQCD calculations [13] with EPS09 parametrization [12] for the nuclear modification of the PDFs in p–Pb collisions. R_{pPb} of heavy-flavour decay muons is compatible with unity within the uncertainties and calculations describe the data within uncertainties in different rapidities. R_{pPb} at low p_T in the backward rapidity region is slightly larger than unity, but consistent with the model calculation.

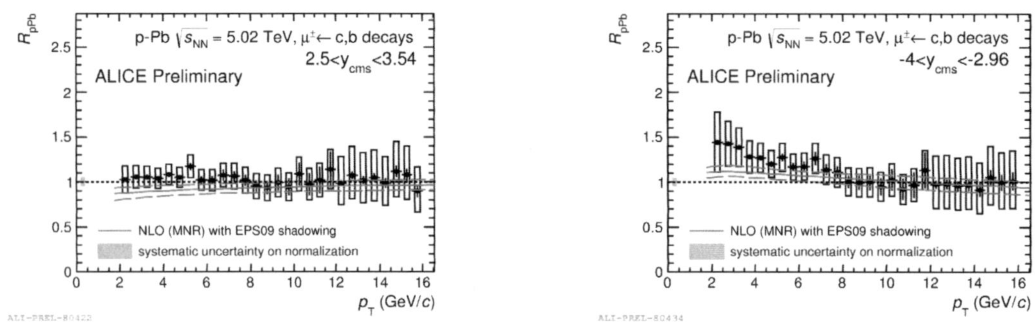

Figure 2. R_{pPb} of heavy-flavour decay muons as a function of p_T at $\sqrt{s_{NN}}$ = 5.02 TeV at forward rapidity, $2.5 < y_{CMS} < 3.54$, (left panel) and backward rapidity, $-4 < y_{CMS} < -2.96$ (right panel). Results are compared to NLO(MNR) pQCD calculations [13] with EPS09 parametrization of shadowing [12].

Figure 3 (left panel) shows the nuclear modification factors (R_{AA}) measured for electrons at mid rapidity ($|y| < 0.6$) and muons at forward rapidity ($2.5 < y < 4$), both from heavy-flavour hadron decays, as a function of p_T in Pb–Pb collisions at $\sqrt{s_{NN}}$ = 2.76 TeV in the 0-10% centrality class. A strong suppression of heavy-flavour decay leptons for $p_T > 3$ GeV/c is observed in central Pb–Pb collisions. Since the measured nuclear modification factors of heavy-flavour decay leptons in p–Pb collisions are compatible with unity, this suppression suggests that heavy quarks lose energy within the created medium. Figure 3 (central panel) shows R_{AA} measured for electrons and muons from heavy-flavour hadron decays as a function of p_T in Pb–Pb collisions at $\sqrt{s_{NN}}$ = 2.76 TeV in less central collisions (muons in the 40-80% and electrons in the 40-50% centrality class). A decrease of suppression is observed with respect to central collisions, which is consistent with the path-length dependence of the parton energy loss in the medium as well as the dependence of the medium density with the collision centrality. Figure 3 (right panel) shows R_{AA} measured for electrons from beauty-hadron decays at mid rapidity ($|y| < 0.8$) as a function of p_T in Pb–Pb collisions at $\sqrt{s_{NN}}$ = 2.76 TeV in the 0-20% centrality class. The observed suppression of electrons from beauty-hadron decay for $p_T > 3$ GeV/c provides a hint of beauty quark in-medium energy loss.

3.2. Elliptic azimuthal anisotropy

The panels of Fig. 4 show the elliptic azimuthal anisotropy (v_2) measured for electrons at mid rapidity ($|y| < 0.7$) and muons at forward rapidity ($2.5 < y < 4$), both from heavy-flavour hadron decays, as a function of p_T in Pb–Pb collisions at $\sqrt{s_{NN}}$ = 2.76 TeV for the centrality classes 0-10% (left panel), 10-20% (central panel) and 20-40% (right panel). A positive v_2 is observed at low p_T, in particular with more than 3σ significance in the range $2 < p_T < 3$ GeV/c for heavy-flavour decay electrons and in the range $3 < p_T < 5.5$ GeV/c for heavy-flavour decay muons in the 20-40% centrality class, which indicates that charm quarks participate in the collective motion of the medium. Results suggest an increase of v_2 from central to semi-central collisions, which is consistent with the expectation of the evolution of the geometrical anisotropy of the overlap region of the colliding nuclei with the centrality.

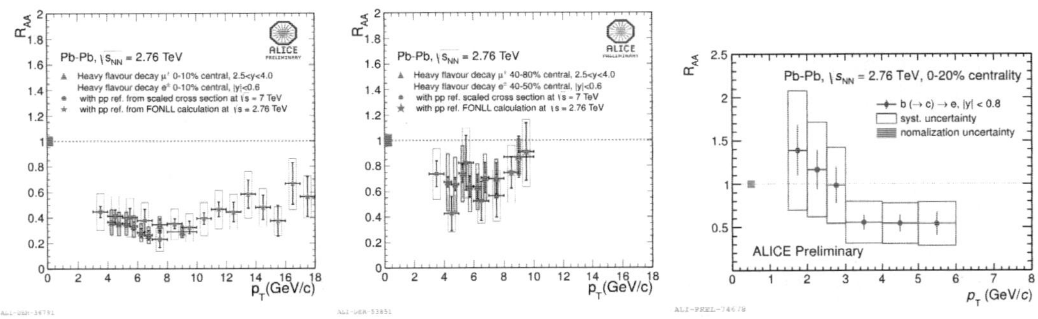

Figure 3. Left panel: R_{AA} of electrons (circle and star symbols) and muons (triangle symbols) from heavy-flavour hadron decays as a function of p_T in Pb–Pb collisions at $\sqrt{s_{NN}}$ = 2.76 TeV for the centrality class 0-10%. Central panel: R_{AA} of electrons measured in the 40-80% and muons in the 40-50% centrality classes. For both panels, electrons are measured at mid rapidity ($|y| < 0.6$), while muons are measured at forward rapidity ($2.5 < y < 4$). Right panel: R_{AA} of beauty-hadron decay electrons as a function of p_T in 0-20% central Pb–Pb collisions at $\sqrt{s_{NN}}$ = 2.76 TeV at mid rapidity ($|y| < 0.8$).

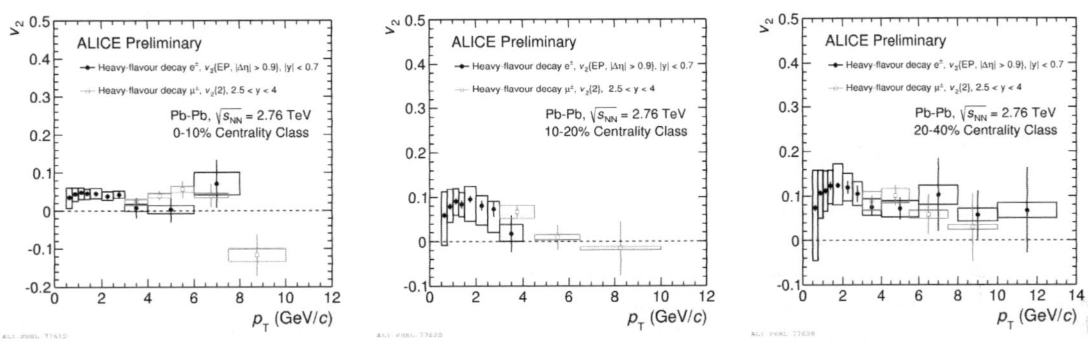

Figure 4. v_2 measured for electrons from heavy-flavour hadron decays at mid rapidity ($|y| < 0.7$) and muons from heavy-flavour hadron decays at forward rapidity ($2.5 < y < 4$) as a function of p_T in Pb–Pb collisions at $\sqrt{s_{NN}}$ = 2.76 TeV in 0-10% (left panel), 10-20% (central panel) and 20-40% (right panel) centrality classes.

Figure 5 shows v_2 (left panel) and R_{AA} (right panel) of electrons from heavy-flavour hadron decays as a function of p_T measured in Pb–Pb collisions in 20-40% and 0-10% centrality classes, respectively [14]. The simultaneous comparison of R_{AA} and v_2 to model calculations [15, 16, 17, 18] presents a challenge for the latter and will eventually allow to constrain model parameters.

3.3. Correlations

Figure 6 (left panel) shows the azimuthal correlation between electrons from heavy-flavour hadron decays and charged hadrons in pp collisions at $\sqrt{s} = 2.76$ TeV at mid rapidity in two different electron p_T ranges [19]. The width of the near-side peak is larger for electrons from beauty-hadron decays than from charm-hadron decays due to the different decay kinematics of charm and beauty hadrons. The measured correlation between heavy-flavour hadron decays and charged hadrons ($e_{HF} - h$) can be expressed as a sum of the correlation between beauty-hadron decay electrons and charged hadrons ($e_b - h$) and the

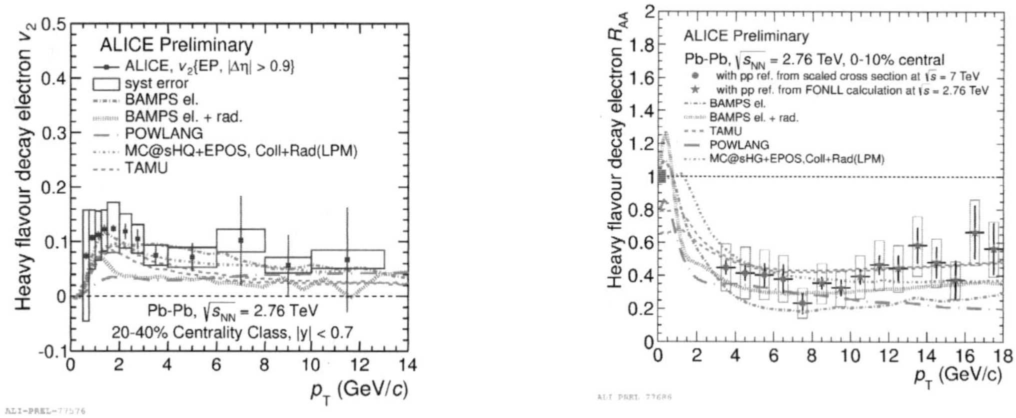

Figure 5. Left panel: v_2 of electrons from heavy-flavour hadron decays as a function of p_T measured in Pb–Pb collisions in mid rapidity ($|y| < 0.7$) in the 20-40% centrality class. Right panel: R_{AA} of electrons from heavy-flavour hadron decays as a function of p_T measured in 0-10% Pb–Pb collisions at mid rapidity ($|y| < 0.6$). For both panels, results are compared to parton transport models [15, 16, 17, 18].

correlation between charm-hadron decay electrons and charged hadrons ($e_c - h$):

$$C(\Delta\varphi,\Delta\eta)_{e_{HF}-h} = A + r_b C(\Delta\varphi,\Delta\eta)_{e_b-h} + (1-r_b)C(\Delta\varphi,\Delta\eta)_{e_c-h}, \tag{4}$$

where $C(\Delta\varphi,\Delta\eta)$ is defined in Eq. 3, A is a constant related to the uncorrelated background, and r_b is the relative contribution of electrons from beauty-hadron decays to the yield of heavy-flavour decay electrons. Equation 4 is used to fit the correlation-distribution templates obtained from a MC simulation with Pythia, for electrons from beauty and charm hadron decays separately, to the measured correlation in order to obtain the parameter r_b. An example of fit is shown in the left panel of Fig. 6. The right panel of Fig. 6 shows r_b and the p_T-differential production cross section of electrons from beauty-hadron decays in pp collisions at $\sqrt{s}$ = 2.76 TeV at mid rapidity ($|y| < 0.7$) obtained via the correlation between electrons from heavy-flavour hadron decays and charged hadrons [19]. The cross section is obtained using the r_b parameter and the measured heavy-flavour decay electron cross section from [20]. Results are compared to the ones obtained using the impact parameter method. The measured relative beauty contribution to the heavy-flavour decay electron yield is described by predictions from different pQCD calculations [8, 21, 22].

Figure 7 (left panel) shows the angular correlation between electrons from heavy-flavour hadron decays and charged particles at mid rapidity ($|\eta| < 0.9$) in pp collisions at $\sqrt{s}$ = 7 TeV and in p–Pb collisions at $\sqrt{s_{NN}}$ =5.02 TeV for three multiplicity classes (0-20%, 20-60%, and 60-100%), which are based on V0A signal amplitude. The p_T ranges of heavy-flavour decay electrons and hadrons are 1-2 GeV/c and 0.5-2 GeV/c, respectively. A similar shape of the angular correlation distributions is observed in low-multiplicity p–Pb collisions at $\sqrt{s_{NN}}$ =5.02 TeV and in pp collisions at $\sqrt{s}$ = 7 TeV. However an increase in the yield in the near and away sides is observed in p–Pb collisions with increasing multiplicity. The observed modification can be quantified by the near side I_{CP}, which is the ratio between the per-trigger yield in high and low multiplicity collisions:

$$I_{CP} = \frac{\text{yield}(-\pi/2 < \Delta\varphi < \pi/2)_{0\%-20\%}}{\text{yield}(-\pi/2 < \Delta\varphi < \pi/2)_{60\%-100\%}}. \tag{5}$$

The measured near side I_{CP} in p–Pb collisions at $\sqrt{s_{NN}}$ =5.02 TeV as a function of electron p_T (central panel of Fig. 7) indicates that there is an enhancement of associated particles for low-p_T electrons in

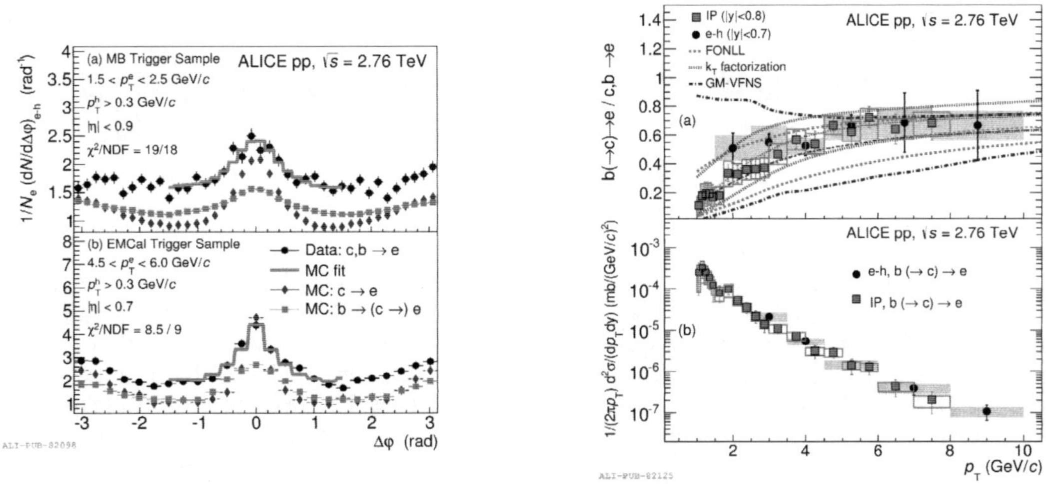

Figure 6. Left panel: azimuthal correlation between electrons from heavy-flavour hadron decays and charged hadrons in pp collisions at $\sqrt{s}$ = 2.76 TeV (a) low-p_T electrons ($1.5 < p^e_T < 2.5$ GeV/c) are measured in the MB trigger sample for $|\eta| < 0.9$ (b) high-p_T electrons ($4.5 < p^e_T < 6$ GeV/c) are measured in the EMCal trigger sample for $|\eta| < 0.7$. The circles indicate the data points, the diamond symbols represent the MC distribution for charm-hadron decay electrons, square symbols are the MC distribution for beauty-hadron decay electrons, and the line represents the MC fit (Eq. 4). Right panel: (a) relative contribution of electrons from beauty-hadron decays to the heavy-flavour decay electron yield. Results are compared to different pQCD calculations [8, 21, 22]. (b) p_T-differential inclusive production cross section of electrons from beauty-hadron decays. The circle symbols represent the data points measured via the azimuthal correlation between electrons from heavy-flavour hadron decays and charged hadrons, and the square symbols are the data points obtained using the impact parameter method.

high-multiplicity p–Pb collisions. The enhancement observed for heavy-flavour decay electrons with $1 < p_T < 2$ GeV/c is further investigated by subtracting the measured correlation in $\Delta\varphi$ and $\Delta\eta$ in low multiplicity (60-100%) collisions from the one measured in high multiplicity (0-20%) events, in order to remove the contribution from jets. The result shown in Fig. 7 (right panel) indicates a long-range correlation in $\Delta\eta$, also observed for light-flavour hadron-hadron angular correlations measured in p–Pb collisions at $\sqrt{s_{NN}}$ = 5.02 TeV [23, 24]. For light-flavour hadrons, this double-ridge structure can be described by calculations including gluon saturation in the initial state [25] and by models assuming an hydrodynamic expansion of final state particles in small systems [26, 27].

4. Conclusions

The p_T-differential production cross section of electrons from beauty-hadron decays and the relative contributions of electrons from beauty-hadron decays to the yield of heavy-flavour decay electrons are obtained in pp collisions at $\sqrt{s}$ = 2.76 TeV using two different techniques, which agree with each other within the uncertainties. Measurements of R_{pPb} of electrons from heavy-flavour hadron decays and from beauty-hadron decays at mid rapidity and R_{pPb} of heavy-flavour decay muons at forward and backward rapidities are compatible with unity. A double-ridge structure is observed in the correlation between heavy-flavour decay electrons and charged particles in p–Pb collisions at $\sqrt{s_{NN}}$ = 5.02 TeV. A similar structure is also observed in light-flavour hadron-hadron correlations, where it can be described by either initial or final state effects. The investigation of the double-ridge structure in heavy-flavour hadron decays, whose statistics will be significantly increased in Run 2, can contribute to clarify the origin of

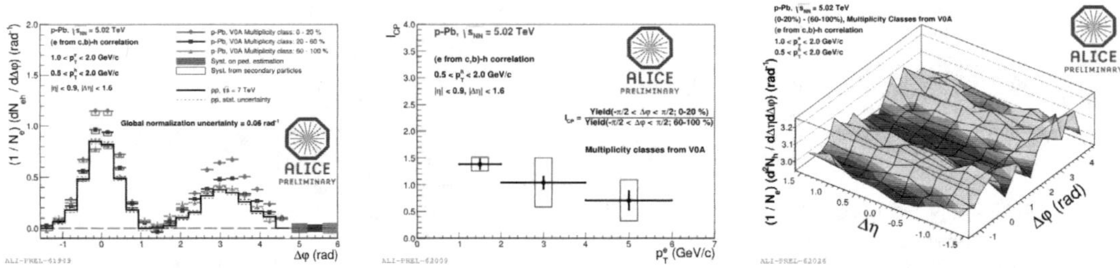

Figure 7. Left panel: azimuthal correlation between low-p_T electrons ($1 < p_T^e < 2$ GeV/c) from heavy-flavour hadron decays and low-p_T charged hadrons ($0.5 < p_T^h < 2$ GeV/c) at mid rapidity ($|\eta| < 0.9$, $|\Delta\eta| < 1.6$). The histogram represents the measument in pp collisions at $\sqrt{s}$ = 7 TeV. The circles, squares and triangles represent the measurement in p–Pb collisions at $\sqrt{s_{NN}}$ = 5.02 TeV in the 0-20%, 20-60% and 60-100% multiplicity classes, respectively. Central panel: ratio between the near-side yields in high (0-20%) and low (60-100%) multiplicity classes (Eq. 5) as a function of electron p_T in p–Pb collisions at $\sqrt{s_{NN}}$ = 5.02 TeV at mid rapidity ($|\eta| < 0.9$, $|\Delta\eta| < 1.6$). Right panel: Difference between the two-particle correlation distributions in $\Delta\varphi$ and in $\Delta\eta$ measured in high (0-20%) and low (60-100%) multiplicity classes in p–Pb collisions at $\sqrt{s_{NN}}$ = 5.02 TeV at mid rapidity ($|\eta| < 0.9$, $|\Delta\eta| < 1.6$).

the effect. Measurements of R_{AA} of heavy-flavour decay leptons in central Pb–Pb collisions at $\sqrt{s_{NN}}$ = 2.76 TeV show a strong suppression of heavy-flavour decay leptons at intermediate and high p_T. Results of R_{pPb} and R_{AA} of heavy-flavour decay leptons suggest small cold nuclear matter effects and indicate that the suppression observed in Pb–Pb collisions is due to the presence of the hot and dense medium. The measurement of the R_{AA} of beauty-hadron decay electrons in the 20% most central Pb–Pb collisions hints at a suppression for $p_T > 3$ GeV/c, which could indicate in-medium energy loss of beauty quarks. A positive v_2 of electrons and muons from heavy-flavour hadron decays is observed with more than 3σ significance in Pb–Pb collisions at $\sqrt{s_{NN}}$ = 2.76 TeV in the 20-40% centrality class, which suggests collective motion of low-p_T charm quarks.

References

[1] M. L. Miller, K. Reygers, S. J. Sanders and P. Steinberg, Ann. Rev. Nucl. Part. Sci. 57, 205 (2007).
[2] A. M. Poskanzer and S. A. Voloshin, Phys. Rev. C 58 (1998) 1671–1678.
[3] ALICE Collaboration, JINST 3 (2008) S08002.
[4] ALICE Collaboration, Phys. Rev. Lett. 106 (2011) 032301.
[5] ALICE Collaboration, Phys. Rev. D 86 (2012) 112007.
[6] CMS Collaboration, CMS-PAS-HIN-12-017.
[7] R. Averbeck, N. Bastid, Z. Conesa del Valle, P. Crochet, A. Dainese, X. Zhang, arXiv:1107.3243 [hep-ph] (2011).
[8] M. Cacciari, M. Greco and P. Nason, JHEP 9805 (1998) 7 ; M. Cacciari, S. Frixione and P. Nason, JHEP 0103 (2001) 6.
[9] ALICE Collaboration, Physics Letters B 721 (2013) 13–23.
[10] ALICE Collaboration, Phys. Rev. Lett. 109 (2012) 112301.
[11] ALICE Collaboration, Physics Letters B 708 (2012) 265–275.
[12] K.J. Eskola, H. Paukkunen, and C.A. Salgado, JHEP04 2009 (2009).
[13] M. Mangano, P. Nason, G. Ridolfi, Nuclear Physics B 373 (1992) 295–345.
[14] Z. C. del Valle for the ALICE Collaboration, Nuclear Physics A 904–905 (2013) 178c–185c.
[15] J. Uphoff, O. Fochler, Z. Xu and C. Greiner, Physics Letters B 717 (2012) 430–435.
[16] M. He, R. J. Fries and R. Rapp, arXiv:1208.0256 [nucl-th] (2012).
[17] M. Monteno, W. M. Alberico, A. Beraudo, A. De Pace, A. Molinari, M. Nardi and F Prino, J. Phys. G. 38 (2011) 124144.
[18] P. B. Gossiaux, R. Bierkandt and J. Aichelin, Phys. Rev. C 79 (2009) 044906.
[19] ALICE Collaboration, arXiv:1405.4144 [nucl-ex] (2014).
[20] ALICE Collaboration, arXiv:1405.4117 [nucl-ex] (2014).
[21] P. Bolzoni, G. Kramer, Nucl.Phys. B 872 (2013) 253.
[22] R. Maciula, A. Szczurek, Phys.Rev. D 87 9 (2013) 094022.

[23] ALICE Collaboration, Phys.Lett. B 719 (2013) 29–41.
[24] ALICE Collaboration, Phys.Lett. B 726 (2013) 164–177.
[25] K. Dusling and R. Venugopalan, Phys. Rev. D 87 (2013) 094034.
[26] K. Werner , I. Karpenko and T. Pierog, Phys. Rev. Lett. 106 (2011) 122004.
[27] P. Bozek and W. Broniowski, arXiv:1403.6042 (2014).

IOP Conf. Series: Journal of Physics: Conf. Series **805** (2017) 012002 doi:10.1088/1742-6596/805/1/012002

Charm and prompt photon production with EPOS

B Guiot[1,2], Iu Karpenko[3], T Pierog[4], G Sophys[2] and K Werner[2]
[1] University Técnica Frederico Santa Maria (UTFSM), Valparaiso, Chile
[2] SUBATECH, University of Nantes - IN2P3/CNRS - EMN, Nantes, France
[3] INFN - Sezione di Firenze, Via G. Sansone 1, I-50019 Sesto Fiorentino (Firenze), Italy
[4] Karlsruhe Institute of Technology (KIT), IKP, Karlsruhe, Germany

E-mail: benjamin.guiot@usm.cl

Abstract. The implementation of heavy quarks and prompt photons in EPOS is presented. One of the interests of event generators is their capability to reproduce exclusive observables. Our results for charms and prompt photons, in p-p collisions at $\sqrt{s} = 7$ TeV, are compared with experiments and FONLL calculations [1].

1. Introduction
An event generator is an important tool for the understanding of heavy ion collisions because it can be used for realistic studies. With EPOS, one of our goals is the study of the quark-gluon plasma. Several results have already been obtained for light flavors in p-Pb [2, 3] and Pb-Pb [4] collisions. In particular, Epos reproduces the ridge structure and the mass splitting of the elliptical flow coefficients v_2 for pions, kaons, and protons in p-Pb collisions.

Recently, heavy quarks and prompt photons have been implemented in EPOS. These "clean" probes are ideal, for instance, for the study of energy loss in the medium. We plan to study heavy quark correlations in Pb-Pb collisions, which could shed light on energy loss mechanisms [5]. This is a project done in collaboration with J. Aichelin, P.B. Gossiaux, M. Nahrgang and Vitallii Ozvenchuk. Heavy quark propagation in the medium will be studied with the transport code MC@sHQ. Realistic initial conditions for the medium and heavy quark distribution will be provided by EPOS.

EPOS using a unified formalism for p-p, p-A and A-A collisions, heavy quarks and prompt photons have to be implemented only for p-p. After a short presentation of this event generator, we will give some details on hard probes production. Finally, results for D mesons and prompt photons will be shown.

2. EPOS and cut Pomerons
We will not do a general presentation of EPOS which can be find in [3]. I will focus on EPOS multiple interaction (MPI) formalism, in particular on cut Pomerons which are important for inelastic processes. Our formalism, called the parton-based Gribov-Regge theory, is described in very details in [6]. It is based on the Gribov-Regge theory where multiple exchange of Pomerons has been introduced to solve an unitarity issue (see [7] for a short review). In this theory, the

IOP Conf. Series: Journal of Physics: Conf. Series **805** (2017) 012002 doi:10.1088/1742-6596/805/1/012002

total cross section (related to the elastic cross section by the optical theorem) can be written as

$$\sigma_{tot} = \sum_{m=1}^{\infty} \sigma_m \tag{1}$$

Here, σ_m, the cross section for the diagram with m cut Pomerons, is associated to an inelastic process.

EPOS formalism is based on the same idea. Energy conservation has been added and the nature of Pomerons is not the same since ours include a pQCD part, figure 1.

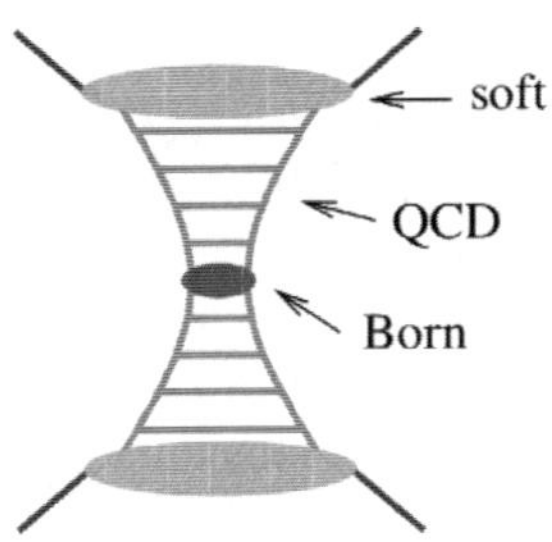

Figure 1. Uncut Pomeron in EPOS.

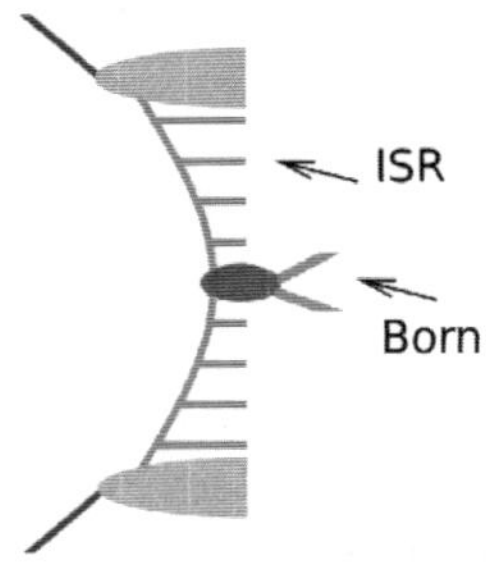

Figure 2. Cut Pomeron in EPOS. Particle production via initial state radiations (ISR) and the born process.

Diagrammatically, our main formula is shown figure 3 and is a generalization of eq. (1) . Cut

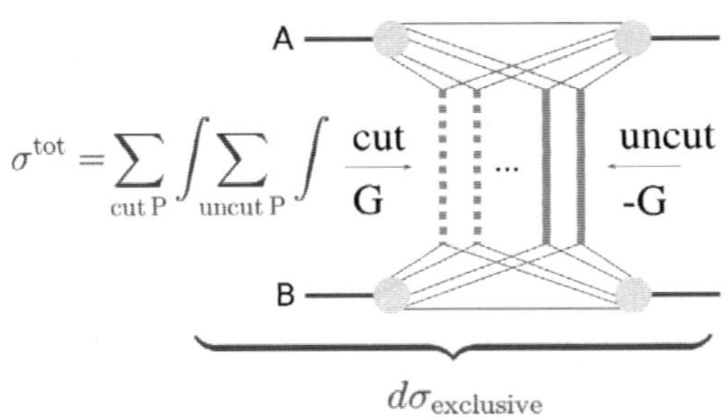

Figure 3. Pink lines : Pomerons. A and B are nuclei. Small horizontal lines are remnants.

Pomerons (figure 2) are central objects for our formalism. First, because they contribute to particle production. In particular, we can write :

$$Multiplicity \propto m \tag{2}$$

$$N_{Dmeson} \propto m \tag{3}$$

m being the number of cut Pomerons. Consequently, we expect a linear rise of the number of D mesons with the multiplicity of charged particles. This has been recently observed by the ALICE collaboration [8] and their results are compared to our simulations figure 4.

Second, because they give initial conditions for the medium. Cut Pomerons are identfied as color strings. Regions with a high density of strings will constitute the core, whereas regions

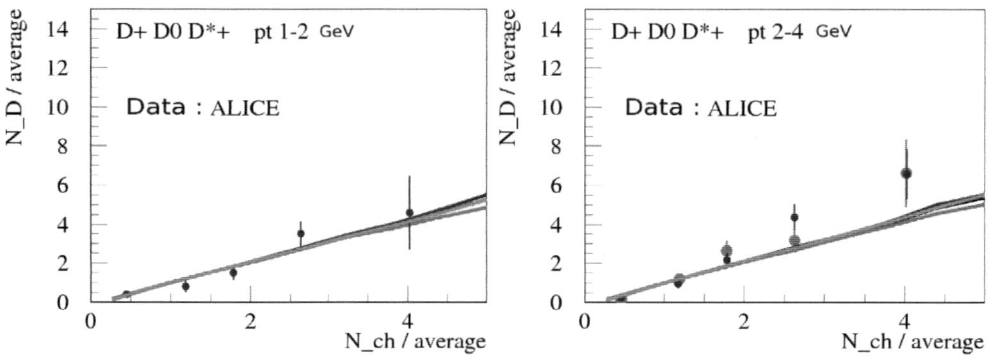

Figure 4. D meson multiplicity as a function of charged particle multiplicity. With EPOS, we obtain a linear rise due to our treatment of multiple interactions. This is in good agreement with ALICE data [8].

with smaller densities will constitute the corona. Quantities like initial energy density for the plasma are provided by cut Pomerons properties. In figure 5 we show the transverse plan with 8 cut Pomerons for one p-Pb collision.

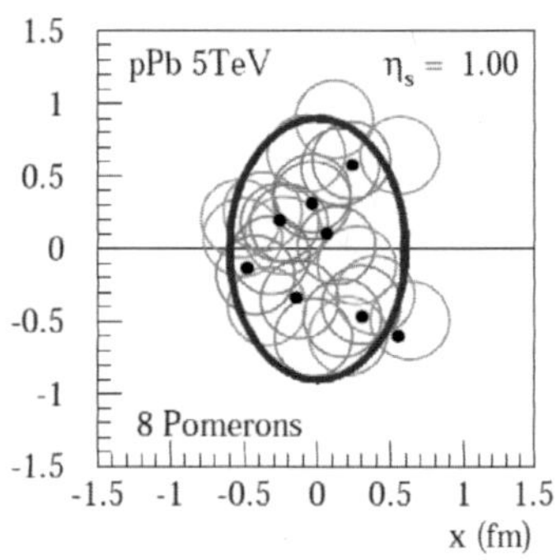

Figure 5. Example of a semi-peripheral p-Pb scattering, with 8 cut Pomerons, showing the transverse plane at space-time rapidity η_s=1. The positions of the Pomerons are indicated by the black dots. String segment having enough energy to escape (corona) are marked green, the red ones constitute the core. The blue line shows that we obtained an elliptical shape for the core.

In the following, I will present the cut Pomeron building, which includes as a special case hard probes production.

3. Cut Pomeron building and hard probes production

A cut Pomeron is shown 2. The blue line represents a parton coming from the nucleus. If its virtuality is smaller than the saturation scale, $Q^2 < Q_s^2$, it will undergo a soft (non-perturbative) evolution, see [6]. We use the saturation scale to take into account (in a phenomenological way) non linear effects at small x. The usual definition is used :

$$Q_s^2 \sim \frac{A^{1/3}}{x^\lambda} \sim \frac{N_{part}}{x^\lambda} \tag{4}$$

N_{part} being computed at the partonic level (see [3] for more details). When the virtuality becomes bigger than the saturation scale, there is an hard evolution based on DGLAP formalism,

IOP Conf. Series: Journal of Physics: Conf. Series **805** (2017) 012002 doi:10.1088/1742-6596/805/1/012002

the so-called (spacelike) parton shower (or spacelike cascade). Partons emitted during this cascade are timelike (horizontal pink lines, figure 2) and their transverse momentum is given by :

$$p_t^2 = (1-\xi)Q^2 - \xi m^2 \tag{5}$$

ξ, the light cone momentum fraction, and m the mass of the timelike parton are shown figure 6.

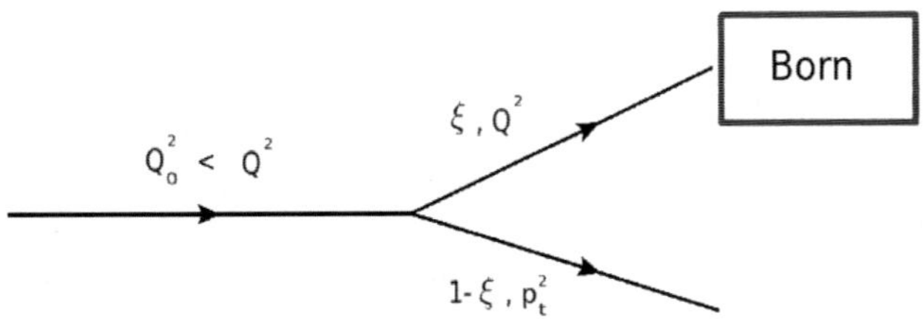

Figure 6. The virtuality , Q^2, increases after each splitting. Partons emitted during this cascade (initial state radiations), are timelike and carry a light cone momentum fraction $(1-\xi)$.

During this cascade, hard probes can be produced via the splittings $g \to c\bar{c}$ and $q \to q\gamma$.

After the spacelike cascade, the born process is done using leading order pQCD cross sections. The factorization scale is :

$$M_F^2 = \frac{q_t^2}{4} \tag{6}$$

with q_t, the transverse momentum for out born partons. The Monte-Carlo for q_t is based on :

$$\frac{d\sigma^{ij}_{born}}{dq_t^2}(\hat{s}, q_t^2) = \frac{1}{16\pi\hat{s}^2\sqrt{1-4q_t^2/\hat{s}}} \sum_{k,l} |M^{ij\to kl}(\hat{s}, q_t^2)|^2 \tag{7}$$

i and j being the flavor of incoming partons and $|M^{ij\to kl}(\hat{s}, q_t^2)|^2$ the squared matrix elements. At LHC energies, the dominant process for heavy flavor production is $gg \to Q\overline{Q}$. Photons are mainly produced by $q(\overline{q})g \to q(\overline{q})\gamma$ and $q\overline{q} \to g\gamma$.

Finally, out born partons and those emitted during the hard evolution have a finite virtuality. Each of them (except photons) will initiate a timelike cascade, as shown figure 7. This is the usual Monte-Carlo resummation of soft and collinear divergences. The ordering variable is the virtuality and we use an additional veto for angular ordering. When the decreasing virtuality reaches the lower bound Q^2_{fin}, the cascade is stopped and hadronization by string fragmentation is performed. Because of our common project MC@sHQ+EPOS, we use :

$$Q^2_{fin} = m^2_{charm} = 2.25\,\mathrm{GeV}^2 \tag{8}$$

The initial virtuality for the cascade is :

$$Q^2_{ini} = p_t^2 + m^2 \tag{9}$$

with p_t^2 defined in eq. (5). As in the spacelike cascade, heavy quarks and prompt photons can be produced via $g \to Q\overline{Q}$ and $q \to q\gamma$. The $Q\overline{Q}$ pair is produced at small angle and small transverse momentum. Consequently, the timelike cascade is important for heavy quark distribution at low p_t and heavy quark correlations (gives the pic at $\Delta\phi = 0$).

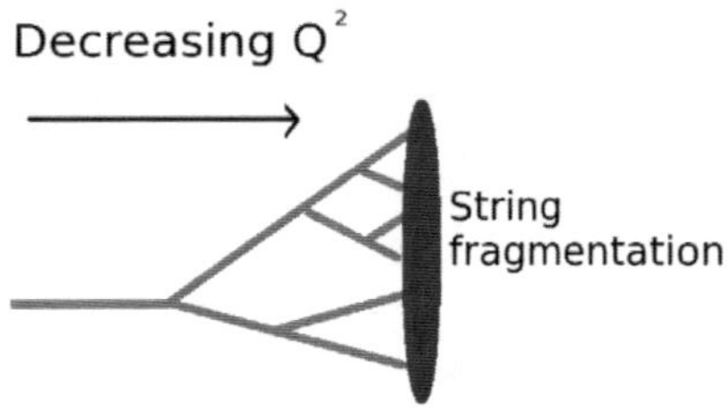

Figure 7. Timelike cascade.

4. Results on charm production

As explained previously, we plan to use charms for the study of energy loss in Pb-Pb collisions. Our charm production has first to be in good agreement with data for p-p collisions. It provides tests for our model, for instance for MPI and parton showers. In figure 8 we show our result for the p_t distribution of charms. A comparison is done with FONLL calculations [1]. p_t distributions

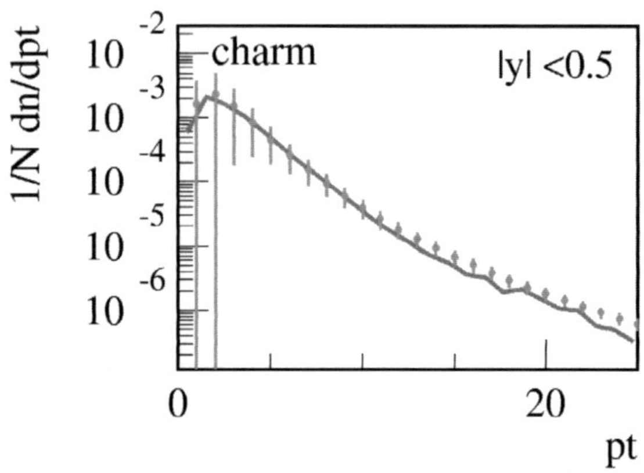

Figure 8. p_t distribution for charms. Green dots are FONLL calculations.

for D mesons are also in good agreement with FONLL calculations and experiments, figures 9 and 10.
Similar results are obtained for D^+, $D0$ and D_s^+ mesons.

5. Isolated photons

Photons can be used for the study of the quark-gluon plasma. For instance, γ/jet events are useful for studies on jet quenching. In this paper, we will study the distribution of the variable x_E :

$$x_E = -\frac{p_t^{asso}}{p_t^{trig}} \cos(\Delta\phi) \tag{10}$$

ideally, p_t^{trig} would be the transverse momentum of a photon produced in the born process. But this is of course a theoretical definition and instead, isolated photons are used in experiments. p_t^{asso} is the transverse momentum of a charged particle in the region $2\pi/3 < \Delta\phi < 4\pi/3$, $0.2\text{GeV} < p_t^{asso} < p_t^{trig}$. $\Delta\phi$ is the azimuthal angle between the photon and its associated particle. The distribution of x_E between 0.2 and 0.9 gives a good approximation of the quark fragmentation function [11]. Comparing distributions of this variable in p-p and Pb-Pb collisions, one can study the modification of the quark fragmentation function by the medium.

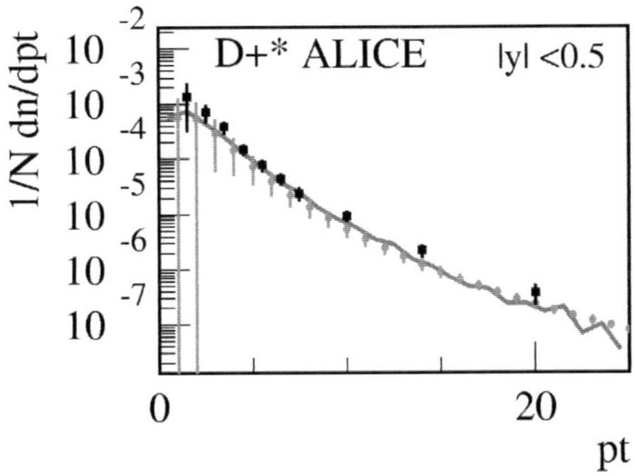

Figure 9. D^{+*} p_t distribution, compared with FONLL calculations (green dots) and ALICE (black squares) [9].

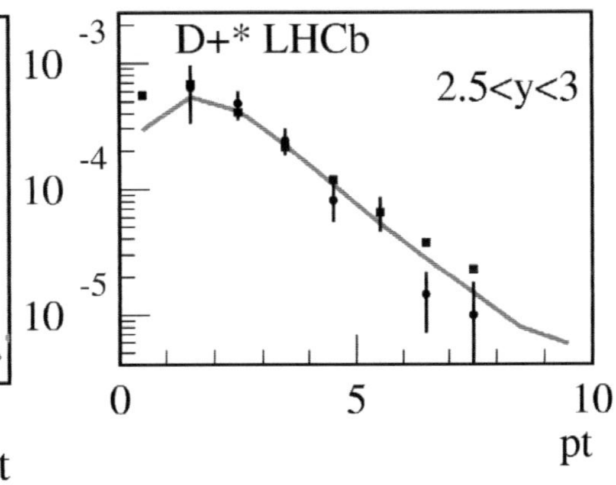

Figure 10. D^{+*} p_t distribution, compared with LHCb (black dots) and Pythia, LHCb tune (black squares) [10].

Recently, fragmentation photons[1] have been implemented in EPOS. Moreover, we use the same isolation criteria as in experiments. Note that this is possible only because, in EPOS, we produce complete events. In each collision, all kind of particles will be produced, even if we are only interested in photons. As expected, fragmentation photons are strongly suppressed by our isolation subroutine, whereas $\sim 98\%$ of direct photons are isolated. Our result for the p_t distribution of isolated photons is shown figure 11 and is in good agreement with CMS measurement [12]. In figure 12 we show the distribution in azimuthal angle of correlations between an isolated

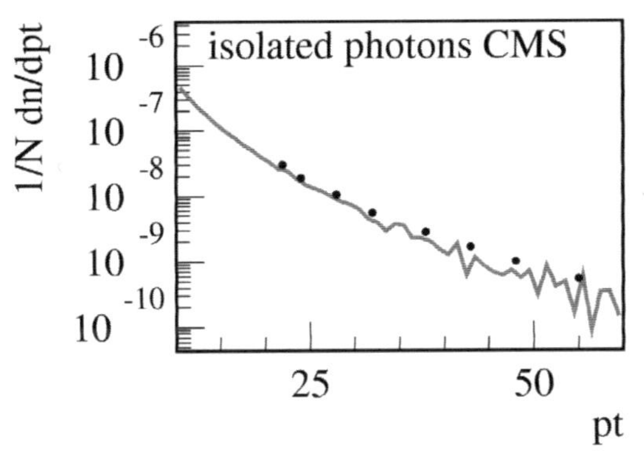

Figure 11. p_t distribution of isolated photons. The comparison is done with CMS [12].

photon and charged particles. By definition, isolated photons are not surrounded by charged particles. This gives the hole ("anti-correlation") at $\Delta\phi = 0$. The fact that we can reproduce the shape of this plot shows the realistic aspect of EPOS events.

Finally, we shows the distribution of the variable x_E figure 13. Our results are in good agreement with experimental data. Here, one can see the interest of event generators which give the

[1] Their are different definitions for fragmentation photons. In this paper, this term is used for photons produced in parton showers.

IOP Conf. Series: Journal of Physics: Conf. Series **805** (2017) 012002 doi:10.1088/1742-6596/805/1/012002

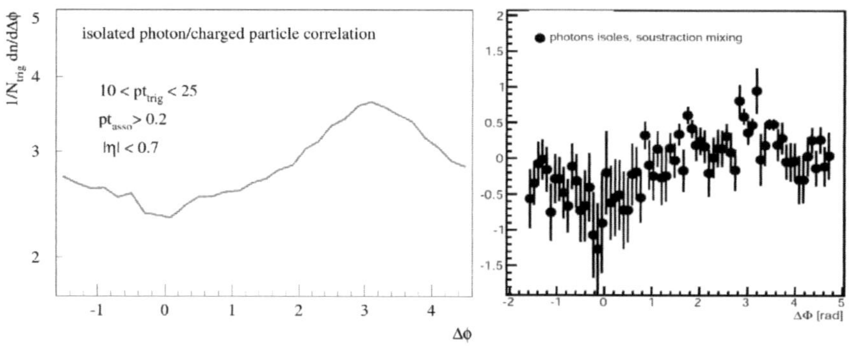

Figure 12. Right : ALICE measurement [13]. Left : EPOS result. The comparison is not done directly with ALICE because we have not performed a mixed event correction. Shapes are comparable, in particular the "anti-correlation" at $\Delta\phi = 0$ is reproduced, showing the realistic implementation of isolation.

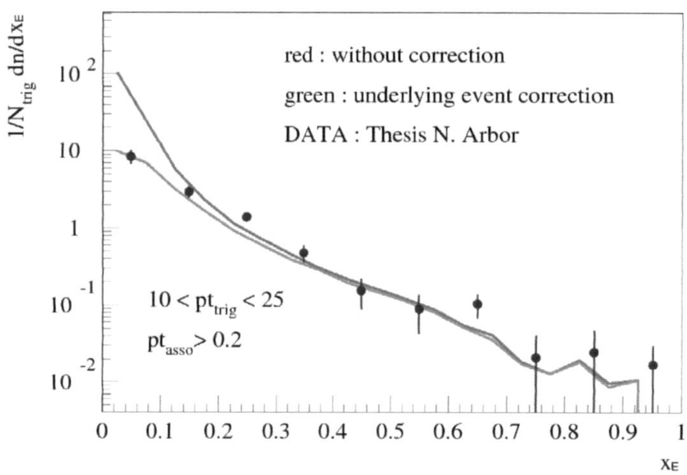

Figure 13. x_E distribution.

possibility to reproduce exclusive observables.

Remark : Fragmentation photons being strongly suppressed by the isolation requirement, these results provide mainly a test for our production of direct photons. Fragmentation photons have still to be tested in more details.

6. Conclusion

Our results for hard probes are in good agreement with experimental data for p-p collisions at 7 TeV. This implementation provides tests for our model, in particular for parton showers. Recently, the bottom quark has also been implemented and some results will be soon published. This event generator shows its good capability to reproduce sophisticated observables like correlations between isolated photons and charged particles and should be a useful tool for the study of the quark-gluon plasma.

Acknowledgement

B Guiot acknowledges the financial support by the TOGETHER project of the Region of "Pays de la Loire". B Guiot gratefully acknowledges generous support from Chilean FONDECYT grants 3160493.

References

[1] Cacciari M, Greco M and Nason P 1998 *arXiv hep-ph/9803400*
[2] Werner K, Bleicher M, Guiot B, Karpenko I and Pierog T 2014 *Phys. Rev. Lett.* **112** 232301
[3] Werner K, Guiot B, Karpenko I and Pierog T 2014 *Phys. Rev.* C **89** 064903
[4] Werner K, Karpenko Iu, Bleicher M, Pierog T, Porteboeuf-Houssais S 2012 *Phys. Rev.* C **85** 064907
[5] Nahrgang M, Aichelin J, Gossiaux P B, Werner K 2013; *arXiv 1310.2218v1*
[6] Drescher H J, Hladik M, Ostapchenko S, Pierog T and Werner K 2001 *Phys. Rep.* **350** 93
[7] Werner K 1993 *Phys. Rep.* **232** 87
[8] Renu Bala, for the ALICE collaboration 2013 *arXiv 1309,6570v1*
[9] Alice collaboration 2012 *arXiv 1111.1553v2*
[10] LHCb-CONF-2010-013
[11] Nicolas Arbor, for the ALICE Collaboration *arXiv 1211.6620v1*
[12] CMS collaboration 2011 *Phys. Rev. Lett.* **106** 082001
[13] Nicolas Arbor, thesis 2013

IOP Conf. Series: Journal of Physics: Conf. Series **805** (2017) 012003 doi:10.1088/1742-6596/805/1/012003

Challenges of direct photon production at forward rapidities and large p_T

Michal Krelina, Jan Cepila
Czech Technical University in Prague, FNSPE, Brehova 7, 11519 Prague, Czech Republic
E-mail: michal.krelina@fjfi.cvut.cz

Jan Nemchik
Czech Technical University in Prague, FNSPE, Brehova 7, 11519 Prague, Czech Republic
Institute of Experimental Physics SAS, Watsonova 47, 04001 Kosice, Slovakia

Abstract. Direct photons produced in interactions with nuclear targets represent a cleaner probe for investigation of nuclear effects than hadrons, since photons have no final state interaction and no energy loss or absorption is expected in the produced hot medium. Therefore, besides the Cronin enhancement at medium-high transverse momenta p_T and isospin effects at larger p_T, one should not expect any nuclear effects. However, this fact is in contrast to the PHENIX data providing an evidence for a significant large-p_T suppression at mid rapidities in central $d + Au$ and $Au + Au$ collisions that cannot be induced by coherent phenomena (gluon shadowing, Color Glass Condensate). We demonstrate that such an unexpected results is subject to deficit of energy induced universally by multiple initial state interactions (ISI) towards the kinematic limits (large Feynman x_F and/or large $x_T = 2p_T/\sqrt{s}$). For this reason, in order to enhance the effects of coherence, one should be cautious going to forward rapidities and higher energies. In the LHC kinematic region ISI corrections are irrelevant at mid rapidities but cause rather strong suppression at forward rapidities and large p_T. Numerical calculations of invariant p_T spectra and the nuclear modification factor were performed within two different models, the color dipole formalism and the model based on k_T-factorization, which are successfully confronted with available data from the RHIC and LHC collider experiments. Finally, we perform also predictions for a strong onset of ISI corrections at forward rapidities and corresponding expected suppression can be verified by the future measurements at LHC.

1. Introduction

Direct photons provide an unique tool to study nuclear effects in proton-nucleus and heavy-ion collisions and represent a cleaner probe than hadron production since they have no final state interactions, either energy loss or absorption in the produced hot medium. For this reason, no convolution with the jet fragmentation function is required and no nuclear effects are expected besides the nuclear shadowing, Cronin enhancement and small isotopic corrections. Thus, direct photons can serve as an additional tool to discriminate between overall nuclear effects and the effects coming from final state interactions typical for strongly interacting particles in heavy-ion collisions. Manifestations of nuclear effects are usually studied through the nucleus-to-nucleon ratio, the so called nuclear modification factor, $R_A(p_T) = \sigma_{pA\to\gamma+X}(p_T)/A\,\sigma_{pp\to\gamma+X}(p_T)$ for pA collisions and $R_{AB}(p_T) = \sigma_{AB\to\gamma+X}(p_T)/AB\,\sigma_{pp\to\gamma+X}(p_T)$ for minimum bias (MB) AB collisions.

IOP Conf. Series: Journal of Physics: Conf. Series **805** (2017) 012003 doi:10.1088/1742-6596/805/1/012003

At medium-high transverse momenta p_T one should take into account the Cronin effect, enhancement of particle production in pA collisions, $R_A(p_T) > 1$. This effect was studied within the color dipole formalism in [1], where the predicted shape and magnitude of the Cronin enhancement were confirmed later by the PHENIX data [2] at RHIC and recently by the ALICE measurements [3] at LHC. However, other models presented in the review [4] were not able to describe successfully the last ALICE data [3].

Since the Cronin enhancement can not be measured precisely by experiments at RHIC and LHC due to difficult identification of direct photons at small and medium-high p_T, in this paper we focused on study of possible nuclear effects in the large-p_T region. In contrast with a naive expectation about an absence of nuclear effects at large p_T the PHENIX data [2] on π^0 production in central dAu collisions provide a clear evidence for a significant suppression at midrapidity, $y = 0$. Such an observation is confirmed also by the PHENIX data on direct photon production in central $AuAu$ collisions [5]. Besides small isotopic corrections, observed attenuation can not be interpreted by the coherence effects (gluon shadowing, color glass condensate) due to large values of Bjorken x.

Alternative interpretation is based on multiple interactions of the projectile hadron and its debris during propagation through the nucleus. The corresponding energy loss is proportional to the hadron energy and the related effects do not disappear at very high energies as was stressed in [6]. In each Fock component the hadron momentum is shared between its constituents: the more constituents are involved, the smaller is the mean energy per parton. This leads to the softer fractional energy distribution of a leading parton, and the projectile parton distribution falls at large $x \to 1$ steeper on a nuclear target than on a proton.

Such softening of the projectile parton fractional energy distribution can be viewed as an effective energy loss of the leading parton due to initial state multiple interactions (ISI). Enhancement of the weight factors for higher Fock states in the projectile hadron with a large number of constituents leads to reduction of the mean fractional energy of the leading parton compared to lower Fock states which dominate the hard reaction on a proton target. Such a reduction is apparently independent of the initial hadron energy and can be treated as an effective loss of energy proportional to the initial hadron energy. A detailed description and interpretation of the corresponding additional suppression was presented also in Refs. [7, 8, 9].

The effect of initial state energy loss (ISI effect) is not effective at high energies and midrapidities. However, it may essentially suppress the cross section approaching the kinematic bound, either in $x_L = 2p_L/\sqrt{s} \to 1$ or $x_T = 2p_T/\sqrt{s} \to 1$ defined at given c.m. energy $\sqrt{s}$. Correspondingly, the proper variable which controls this effect is $\xi = \sqrt{x_L^2 + x_T^2}$.

The magnitude of suppression was evaluated in Ref. [6, 10]. It was found within the Glauber approximation that each interaction in the nucleus leads to a suppression factor $S(\xi) \approx 1 - \xi$. Summing up over the multiple initial state interactions in a pA collision with impact parameter b one arrives at a nuclear ISI-modified parton distribution function (PDF) $F_{a/p}(x, Q^2) \Rightarrow F_{a/p}^{(A)}(x, Q^2, b)$, where

$$f_{a/p}(x, Q^2) \Rightarrow f_{a/p}^{(A)}(x, Q^2, b) = C_v f_{a/p}(x, Q^2) \frac{e^{-\xi \sigma_{eff} T_A(b)} - e^{-\sigma_{eff} T_A(b)}}{(1 - \xi)\left(1 - e^{-\sigma_{eff} T_A(b)}\right)}. \quad (1)$$

Here $\sigma_{eff} = 20$ mb [6] is the effective hadronic cross section controlling the multiple interactions. The normalization factor C_v in Eq. (1) is fixed by the Gottfried sum rule, $T_A(\vec{b})$ is the nuclear thickness function at given impact parameter b normalized to the mass number A. It was found that such an additional nuclear suppression due to the ISI effects represents an energy independent feature common for all known reactions, experimentally studied so far, with any leading particle (hadrons, Drell-Yan dileptons, charmonium, etc.).

IOP Conf. Series: Journal of Physics: Conf. Series **805** (2017) 012003 doi:10.1088/1742-6596/805/1/012003

Using PDFs modified by the ISI energy loss, Eq. (1), one can predict much stronger onset of nuclear suppression in a good agreement with available data from the BRAHMS and STAR [11] experiments at forward rapidities (large x_L) in dA collisions [6, 10]. An alternative interpretation [12] is based on the coherence effects, which should disappear at lower energies because $x \propto 1/\sqrt{s}$ increases. However, according to Eq. (1) the suppression caused by the ISI energy loss scales in Feynman $x_F = x_L$ and should exist at any energy. Thus, by reducing the collision energy one should provide a sensitive test for the models. Expectation of no suppression following from CGC at forward rapidities and small energies is in contradiction with data from the NA49 experiments [13] at SPS obtained at much smaller energy than BRAHMS. This observation confirms an onset of suppression at forward rapidities with entirely interpretation based on the ISI energy loss.

The ISI energy loss also affects the p_T dependence of the nuclear suppression in heavy ion collisions. These effects are calculated similarly to $p(d)A$ collisions using the modified PDFs, Eq. (1), for nucleons in both colliding nuclei.

In order to test theoretical uncertainties, in this paper we calculate p_T-spectra and nuclear suppression of direct photons produced on nuclear targets at RHIC and LHC energy using two different models. Corresponding results obtained within the model based on k_T-factorisation [14] will be compared with the color dipole approach [15].

2. Model based on k_T-factorisation

Here the process of direct photon production to the leading order can be treated as a collision of two hadrons where a quark from one hadron annihilates with an antiquark from the other hadron into a real photon. In vacuum (e.g. in pp collisions), in calculations of the invariant cross section of direct photon production we employ the model proposed in [16]:

$$E\frac{d^3\sigma^{pp\to\gamma X}}{d^3p} = K\sum_{abd}\int d^2k_{Ta}d^2k_{Tb}\frac{dx_a}{x_{Ra}}\frac{dx_b}{x_{Rb}}g_p(k_{Ta},Q^2)\,g_p(k_{Tb},Q^2)$$
$$\times\, F_{a/p}(x_a,Q^2)\,F_{b/p}(x_b,Q^2)\,\frac{\hat{s}}{\pi}\frac{d\hat{\sigma}^{ab\to\gamma d}}{d\hat{t}}\,\delta(\hat{s}+\hat{t}+\hat{u}), \qquad (2)$$

which corresponds to the collinear factorization expression modified by an intrinsic transverse momentum dependence. In Eq. (2) $K \approx 1.0 - 1.5$ is the normalization factor depending on the c.m. energy, $F_{i/p}(x_i,Q^2) = x_i\, f_{i/p}(x_i,Q^2)$ with PDF $f_{i/p}(x_i,Q^2)$, $d\hat{\sigma}/d\hat{t}$ is the cross section of hard parton scattering, x_a, x_b are fractions of longitudinal momenta of the incoming hadrons. The radial variable is defined as $x_{Ri}^2 = x_i^2 + 4k_{Ti}^2/s$, $\hat{s}, \hat{t}, \hat{u}$ are the parton Mandesltam variables and $\vec{k}_{T_i}$ is transverse momentum of parton.

The distribution of the initial parton transverse momentum is described by the Gaussian form [16],

$$g_p(k_T,Q^2) = \frac{1}{\pi\langle k_T^2\rangle_N(Q^2)}e^{-k_T^2/\langle k_T^2\rangle_N(Q^2)} \qquad (3)$$

with the scale dependent parametrization of the mean intrinsic transverse momentum from [16], $\langle k_T^2\rangle_N(Q^2) = \langle k_T^2\rangle_0 + 0.2\,\alpha_S(Q^2)\,Q^2$, where $\langle k_T^2\rangle_0 = 0.2$ GeV^2 and 2.0 GeV^2 for quarks and gluons, respectively.

For the hard parton scattering cross section we use regularization masses $\mu_q = 0.2$ GeV and $\mu_G = 0.8$ GeV for quark and gluon propagators, respectively. In all calculations we take the scale $Q^2 = p_T^2$. For PDFs we used MSTW2008 [17] parametrization.

The differential cross section for direct photon production in $p+A$ and $A+A$ collisions then can be treated as

$$E\frac{d^3\sigma^{pA\to\gamma X}}{d^3p} = \int d^2b\,T_A(\vec{b})E\frac{d^3\tilde{\sigma}^{pp\to\gamma X}}{d^3p} \qquad (4)$$

and

$$E\frac{d^3\sigma^{AB\to\gamma X}}{d^3p} = \int d^2b\, d^2s\, T_A(\vec{s})T_B(\vec{b}-\vec{s})E\frac{d^3\tilde{\sigma}^{pp\to\gamma X}}{d^3p}, \tag{5}$$

respectively.

In Eqs. (4) and (5) the pp-invariant cross section has the same form as is given by Eq.(2) except for a modification of PDFs to nuclear ones (nPDF) $F_{i/A}(b, x_i, Q^2) = R_i^A(x, Q^2)\left(\frac{Z}{A}x_i f_{i/p}(x, Q^2) + \left(1 - \frac{Z}{A}\right) x_i f_{i/n}(x, Q^2)\right)$, where $R_i^A(x, Q^2)$ includes the nuclear shadowing via the nuclear modification factor from EPS09 [18]. Invariant cross section $E\, d^3\tilde{\sigma}^{pp\to\gamma X}/d^3p$ in Eqs. (4) and (5) contains also a nuclear modified distribution of the initial parton transverse momentum as reads,

$$g_A(k_T, Q^2, b) = \frac{1}{\pi\langle k_T^2\rangle_A(Q^2, b)} e^{-k_T^2/\langle k_T^2\rangle_A(Q^2,b)}, \tag{6}$$

where impact parameter dependent variance $\langle k_T^2\rangle_A(Q^2, b) = \langle k_T^2\rangle_N(Q^2) + \Delta k_T^2(b)$ is larger than in pp collisions due to the nuclear k_T-broadening $\Delta k_T^2(x, b) = 2\, C(x)\, T_A(b)$ evaluated within the color dipole formalism [19]. The factor $C(x)$ is related to the dipole cross section $\sigma_{\bar{q}q}$, which describes the interaction of the $\bar{q}q$ pair with a nucleon, as $C(x) = \left.\frac{d\sigma_{\bar{q}q}^N(x,r)}{dr^2}\right|_{r=0}$. Note that for gluons the nuclear broadening is larger due to the Casimir factor 9/4. For the dipole cross section we adopt the GBW parametrization from [20].

3. Color Dipole formalism

The color dipole formalism is treated in the target rest frame, where the process of direct photon production can be viewed as a radiation of a real photon by a projectile quark [15]. Assuming only the lowest $|q\gamma\rangle$ Fock component, the p_T distribution of the photon bremsstrahlung in quark-nucleon interaction can be expressed as a convolution of the dipole cross section $\sigma_{q\bar{q}}^N(\alpha\rho, x)$ and the light-cone (LC) wave functions of the projectile $q + \gamma$ fluctuation $\Psi_{\gamma q}(\alpha, \vec{\rho})$ [15]:

$$\frac{d\sigma(qN\to\gamma X)}{d\ln\alpha d^2p_T} = \frac{1}{(2\pi)^2}\int\sum_{in,f} d^2\rho_1\, d^2\rho_2\, e^{i\vec{p}_T\cdot(\vec{\rho}_1-\vec{\rho}_2)}\,\Psi_{\gamma q}^*(\alpha, \vec{\rho}_1)\,\Psi_{\gamma q}(\alpha, \vec{\rho}_2)\,\Sigma(\alpha, \rho_1, \rho_2, x_2), \tag{7}$$

where $\Sigma(\alpha, \rho_1, \rho_2, x) = \{\sigma_{\bar{q}q}^N(\alpha\rho_1, x) + \sigma_{\bar{q}q}^N(\alpha\rho_2, x) - \sigma_{\bar{q}q}^N(\alpha(\vec{\rho}_1 - \vec{\rho}_2, x))\}/2$.

The differential hadronic cross section for direct photon production in pp collisions can be expressed as a convolution of the differential cross section, Eq. (7) with corresponding PDFs

$$\frac{d^3\sigma^{pp\to\gamma X}}{dx_1 d^2p_T} = \frac{1}{x_1+x_2}\int\frac{d\alpha}{\alpha}\sum_q e_q^2\left(\frac{x_1}{\alpha}\, f_{q/p}\left(\frac{x_1}{\alpha}, Q^2\right) + \frac{x_1}{\alpha}\, f_{\bar{q}/p}\left(\frac{x_1}{\alpha}, Q^2\right)\right)\frac{d\sigma^{qp\to\gamma p}}{d\ln\alpha d^2p_T}, \tag{8}$$

where e_q is a quark charge, $\alpha = p_\gamma^+/p_q^+$ is a fraction of quark LC momenta taken by the photon and Bjorken variables x_1 and x_2 are connected with the Feynman variable as $x_F = x_1 - x_2$ with $x_1 = p_\gamma^+/p_p^+$ in the target rest frame. In all calculation we use the scale $Q^2 = p_T^2$, for PDFs we take the GRV98 parametrization from [21] and for the color dipole cross section we adopt GBW parametrization [20].

Mechanism of direct photon production in pA and AA collisions is controlled by the mean coherence length, $l_c = \left\langle\frac{2E_q\alpha(\alpha-1)}{\alpha^2m_q^2+p_T^2}\right\rangle_\alpha$, where $E_q = x_q s/2m_N$ and $m_q = 0.2$ GeV are the energy and mass of projectile quark, respectively. The variable $x_q = x_1/\alpha$ denotes a fraction of the proton momentum carried by the quark. The onset of nuclear shadowing requires a sufficiently

IOP Conf. Series: Journal of Physics: Conf. Series **805** (2017) 012003 doi:10.1088/1742-6596/805/1/012003

long coherence length (LCL), $l_c \gtrsim R_A$, where R_A is the nuclear radius. This LCL limit can be safely used for the RHIC and LHC kinematic regions especially at forward rapidities and leads to a simple incorporation of shadowing effects via eikonalization of $\sigma^N_{\bar{q}q}(\rho, x)$ [22], i.e. performing the following substitutions in Eq. (7):

$$\sigma^N_{\bar{q}q}(\alpha\rho, x) \Rightarrow \sigma^A_{\bar{q}q}(\alpha\rho, x) = 2 \int d^2s\, \sigma^A_{\bar{q}q}(\vec{s}, \alpha\rho, x) \tag{9}$$

for proton-nucleus interations and

$$\sigma^N_{\bar{q}q}(\alpha\rho, x) \Rightarrow \sigma^{AB}_{\bar{q}q}(\alpha\rho, x) = \int d^2b\, d^2s \left[\sigma^B_{\bar{q}q}(\vec{s}, \alpha\rho, x) T_B(\vec{b} - \vec{s}) + \sigma^A_{\bar{q}q}(\vec{b} - \vec{s}, \alpha\rho, x) T_B(\vec{s})\right] \tag{10}$$

for heavy-ion collisions where

$$\sigma^A_{\bar{q}q}(\vec{s}, \alpha\rho, x) = 1 - \left(1 - \tfrac{1}{2A}\sigma^N_{\bar{q}q}(\alpha\rho, x) T_A(\vec{s})\right)^A . \tag{11}$$

In the LCL limit, besides the lowest $|q\gamma\rangle$ Fock state one should include also higher Fock components containing gluons. They cause an additional suppression, known as the gluon shadowing (GS). Gluon shadowing is incorporated via attenuation factor R_G [23] as the modification of the nuclear thickness function $T_A(\vec{s}) \Rightarrow T_A(\vec{s}) R_G(x_2, Q^2, A, \vec{s})$ in Eq. (9) for $p+A$ interaction and $T_A(\vec{s}) \Rightarrow T_A(\vec{s}) R_G(x_2, Q^2, A, \vec{s})$ and $T_B(\vec{b} - \vec{s}) \Rightarrow T_B(\vec{b} - \vec{s}) R_G(x_1, Q^2, A, \vec{b} - \vec{s})$ in Eq. (10) for heavy-ion collisions.

4. Results

Figs. 1 and 2 show a comparison of both models with PHENIX [24] and CMS [25] data on direct photon production in pp collisions at midrapidity and c.m. energy $\sqrt{s} = 200$ GeV and $\sqrt{s} = 2760$ GeV, respectively For both energies the model based on k_T-factorisation (blue solid lines) and calculations within the color dipole formalism (red solid lines) agree with data reasonably. However, the former describes the data better in the low-p_T region especially at smaller energies due to an absence in this kinematic region of the precise parametrization of the dipole cross section [20] used in calculations. More precise recent parametrization (see [26], for example) improve an agreement with data at small p_T.

Fig. 3 shows a confrontation of the PHENIX data [5] on direct photon production in $AuAu$ collisions with both models. Experimental values were measured at $\sqrt{s} = 200$ GeV and at midrapidity for several centralities $0 - 10\,\%$, $40 - 50\,\%$ and minimum-bias (MB). Blue and red lines represent calculations within the model based on k_T-factorisation and the color dipole formalism, respectively. The solid and dashed lines represent calculations with and without ISI effect. As was mentioned above, at small and medium-high p_T the different shape and magnitude of the Cronin enhancement predicted within both models is caused predominantly by an absence of the precise parametrization of the dipole cross section [20] used in calculations. We expect that more precise recent parametrization [26] of the dipole cross section used in the color dipole formalism leads to a better agreement with the model based on k_T factorization in the small p_T region. Similarly as was demonstrated above, both models agree well with data in the large-p_T region. Moreover, the data on $R_{AuAu}(p_T)$ at centrality $0 - 10\,\%$ indicate a significant suppression at large $p_T \gtrsim 17$ GeV that can not be interpreted by coherence effects. Calculations within both models including ISI corrections clearly demonstrate the observed large-p_T attenuation.

Fig. 4 shows predictions of both models for $R_{AuAu}(p_T)$ at forward rapidity $y = 3$ and for the same centralities that are indicated in Fig 3. Here we predict a strong suppression for all centralities due to ISI effects.

In Fig. 5 we compare predictions of both models with available data from CMS experiment [25] on direct photon production in $PbPb$ collisions at c.m. energy $\sqrt{s} = 2760$ GeV for three

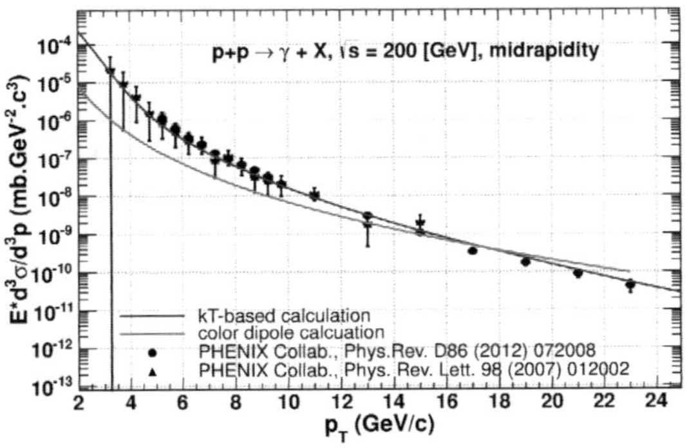

Figure 1. Invariant cross section for direct photon production in pp collisions. The data from the PHENIX experiment [24] are compared with the model based on k_T factorization (blue line) and with calculations based on color dipole formalism (red line).

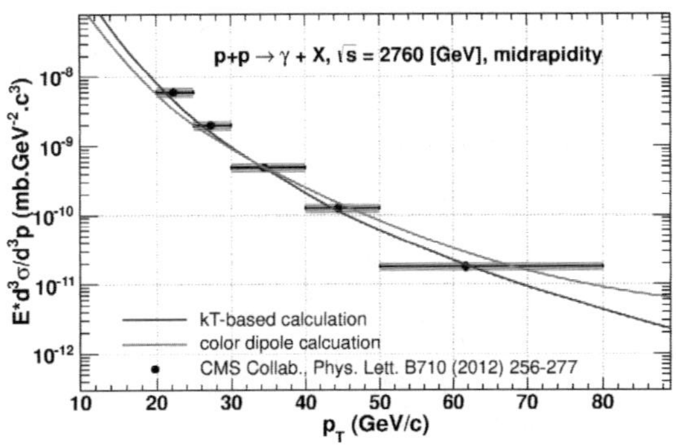

Figure 2. The same as Fig. 1 but with data from CMS experiment [25].

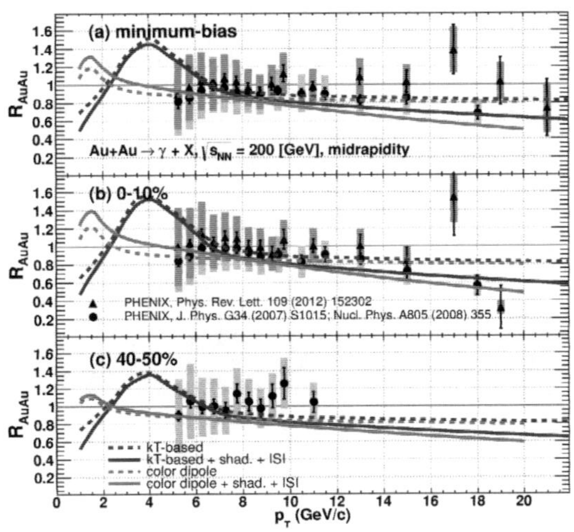

Figure 3. Comparison of the PHENIX data on R_{AuAu} [5] at midrapidity and for several centralities with the model based on k_T factorization (blue line) and with calculations based on the color dipole formalism (red lines). The solid and dashed lines represent calculations with and without ISI effects.

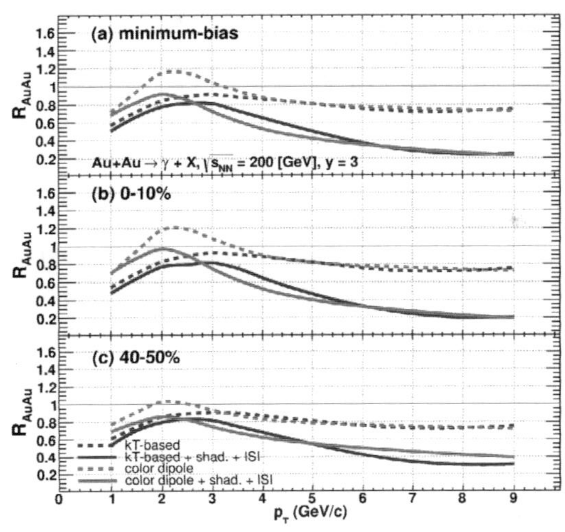

Figure 4. The same as Fig. 3 but at rapidity $y = 3$.

different intervals of centralities. The predictions of both models are qualitatively very close in good accordance with data. They also demonstrate a very weak onset of ISI corrections at large p_T.

Fig. 6 shows predictions from both models for R_{PbPb} at forward rapidity $y = 4$ for the same centrality intervals as are depicted in Fig. 5. We predict a strong suppressions for all centralities due to ISI effects.

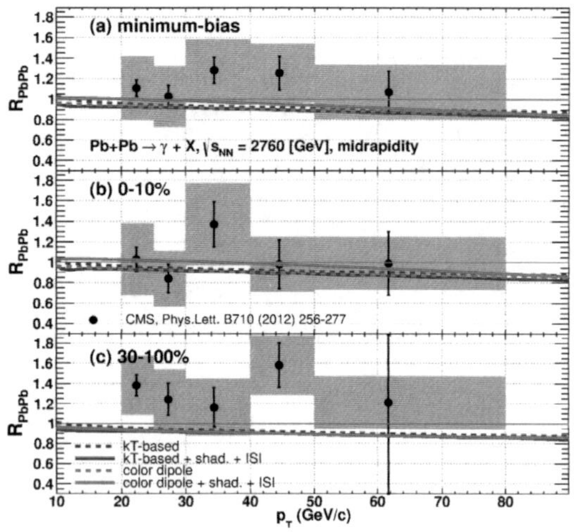

Figure 5. The same as Fig. 3 but with CMS data [25].

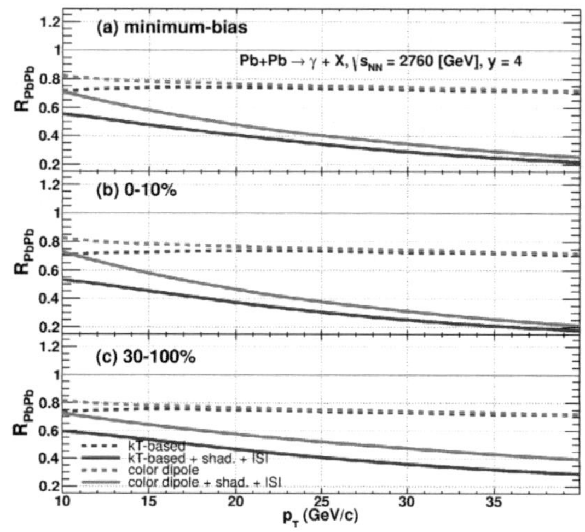

Figure 6. The same as Fig. 4 but at rapidity $y = 4$ and at c.m. energy $\sqrt{s} = 2760$ GeV.

Predictions for $R_{dAu}(p_T)$ from both models are compared in Fig. 7 with PHENIX data [27] on direct photon production in dAu collisions at midrapidity and at $\sqrt{s} = 200$ GeV. Here we predict a sizeable effect of ISI corrections that can be verified in the future by data obtained at very large $p_T \gtrsim 15 - 20$ GeV. The same Fig. 7 also clearly manifests a strong rise of ISI effects with rapidity at fixed p_T values as was discussed in Sect. 1.

Similarly as was mentioned above and presented in Fig. 3 one can see a quantitative difference between both models in predictions of the shape and magnitude of the Cronin enhancement at different rapidities.

Fig. 8 contains predictions for direct photons produced at LHC c.m. energy $\sqrt{s} = 5020$ GeV in pPb collisions at different rapidities $y = 0, 2$ and 4. Here we predict a significant large-p_T suppression due to ISI effects only at rapidities $y \gtrsim 2$. The expected rise of nuclear attenuation with rapidity can be verified in the future by experiments at LHC.

5. Conclusions

We study production of direct photons in pp, $p(d)A$ and AA collisions at RHIC and LHC energies using two different models: the model based on k_T factorization and the model based on the color dipole formalism. The main motivation for a such investigation was to test the theoretical uncertainties in predictions of corresponding variables that can be verified by available data.

Both models describe reasonable well the data on direct photon production in pp collisions. The model based on k_T-factorisation shows a better agreement with data in the low-p_T region. This fact is a consequence of an absence of the more precise determination of the dipole cross section in this kinematic region as that used in calculations within the color dipole formalism.

Investigating direct photon production on nuclear targets, at small and medium-high values of p_T we found a significant difference between predictions of the shape and magnitude of the Cronin enhancement from both models. However, we expect a better agreement between the both models using more precise recent parameterizations of the dipole cross section as is presented in [26], for example. In the large-p_T region we found a good agreement of both models with available data on nuclear modification factors R_A and R_{AA} at RHIC and LHC.

Besides the Cronin enhancement, isospin corrections and coherence effects we investigated also additional suppression due to initial state effective energy loss (ISI effects). We

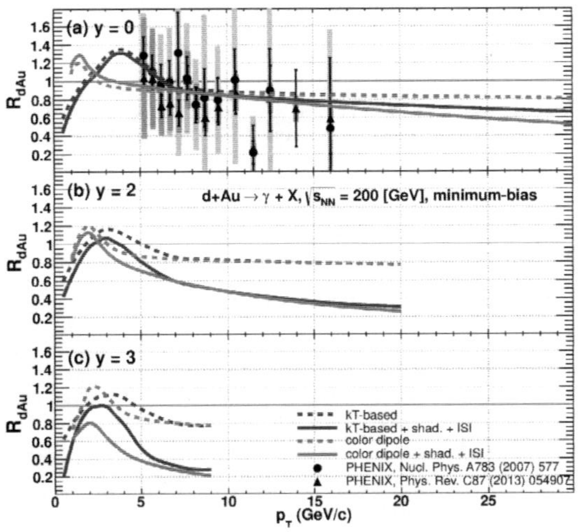

Figure 7. Predictions from both models at different rapidities $y = 0, 2$ and 3 vs. PHENIX data [27] on $R_{dAu}(p_T)$ at midrapity and at $\sqrt{s} = 200$ GeV

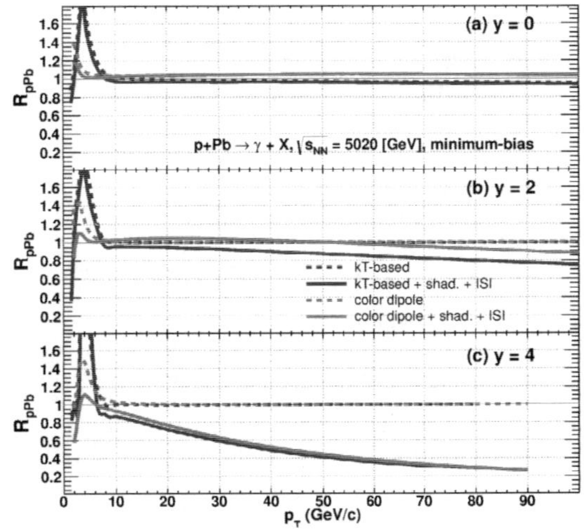

Figure 8. Predictions for R_{pPb} from both models at $\sqrt{s}$ = 5020 GeV and at different rapidities $y = 0, 2$ and 4.

demonstrated that ISI effects cause a strong suppression at forward rapidities and large p_T leading so to breakdown of the QCD factorisation. In the RHIC kinematic region no coherence effects are possible at large p_T. However, the PHENIX data on direct photon production in $AuAu$ interactions clearly indicate a significant large-p_T suppression that can be explained entirely by ISI effects. The ISI effects are practically irrelevant at LHC but we predict a strong nuclear suppression at forward rapidities that can be verified by the future measurements at RHIC and LHC.

Acknowledgements
This work has been supported in part by the grant 13-20841S of the Czech Science Foundation (GAČR), by the Grant MŠMT LG13031, by the Slovak Research and Development Agency APVV-0050-11 and by the Slovak Funding Agency, Grant 2/0020/14 by the European social fund within the framework of realizing the project "Support of inter-sectoral mobility and quality enhancement of research teams at Czech Technical University in Prague", CZ.1.07/2.3.00/30.0034.

References

[1] B. Z. Kopeliovich, et al., Phys. Rev. Lett. **88**, 232303 (2002).
[2] S. S. Adler, et al. (PHENIX Collab.), Phys. Rev. Lett. **98**, 172302 (2007); Phys. Rev. Lett. **109**, 152302 (2012).
[3] B. Abelev, et al. (ALICE Collab.), Phys. Rev. Lett. **110**, 082302 (2013).
[4] J. Albacete, N. Armesto, et al., Int. J. Mod. Phys. E**22**, 1330007 (2013).
[5] S. Afanasiev et al. (PHENIX Collab.), Phys. Rev. Lett. **109**, 152302 (2012); T. Sakaguchi Nucl. Phys. A **805**, 355 (2008).
[6] B Z. Kopeliovich et al., Phys. Rev. C**72**, 054606 (2005); B. Z. Kopeliovich and J. Nemchik, J. Phys. G**38**, 043101 (2011).
[7] J. Nemchik, et al., Phys. Rev. C**78**, 025213 (2008); Nucl. Phys. A**830**, 611c (2009).
[8] B. Z. Kopeliovich and J. Nemchik, Phys. Rev. C**86**, 054904 (2012).
[9] B. Z. Kopeliovich, J. Nemchik, I. K. Potashnikova and I. Schmidt, Int. J. Mod. Phys. E **23**, 1430006 (2014).

[10] B. Z. Kopeliovich and J. Nemchik, J. Phys. G **38**, 043101 (2011).
[11] I. Arsene, et al. (BRAHMS Collab.), Phys. Rev. Lett. **93**, 242303 (2004); J. Adams, et al. (STAR Collab.), Phys. Rev. Lett. **97**, 152302 (2006).
[12] D. Kharzeev, Y.V. Kovchegov and K. Tuchin, Phys. Lett. B **599**, 23 (2004).
[13] B. Boimska [NA49 Collaboration], Ph.D. Dissertation **CERN-THESIS-2004-035**
[14] R. P. Feynman, R. D. Field and G. C. Fox, Phys. Rev. D**18**, 3320 (1978).
[15] B. Z. Kopeliovich, A. V. Tarasov and A. Schafer, Phys. Rev. C**59**, 1609 (1999).
[16] X. N. Wang, Phys. Rev. C**61**, 064910 (2000).
[17] A. D. Martin, W. J. Stirling, R. S. Thorne and G. Watt, Eur. Phys. J. C**63**, 189-285 (2009).
[18] K. J. Eskola, H. Paukkunen and C. .A. Salgado, JHEP **0904**, 065 (2009).
[19] M. B. Johnson, B. Z. Kopeliovich and A. V. Tarasov, Phys. Rev. C**63**, 035203 (2001).
[20] K. Golec-Biernat and M. Wüsthoff, Phys. Rev. D**59**, 014017 (1998).
[21] M. Gluck, E. Reya and A. Vogt, Eur. Phys. J. C**5**, 461 (1998).
[22] B. Z. Kopeliovich, L. I. Lapidus and A. B. Zamolodchikov, JETP Lett. **33**, 595 (1981).
[23] B. Z. Kopeliovich, J. Nemchik, A. Schafer and A. V. Tarasov Phys. Rev. C **65**, 035201 (2002).
[24] A. Adare, et al. (PHENIX Collab.), Phys. Rev. D **86**, 072008 (2012); S. S. Adler et al. (PHENIX Collab.), Phys. Rev. Lett **98**, 012002 (2007).
[25] S. Chatrchyan, et al. (CMS Collab.), Phys. Lett. B **710**, 256-277 (2012).
[26] A. H. Rezaeian, at al., Phys. Rev. D**87**, 034002 (2013).
[27] D. Peressounko (PHENIX Collab.), Nucl. Phys. A **783**, 577-582 (2007); A. Adare, et al. (PHENIX Collab.), Phys. Rev. C **87**, 054907 (2013).

IOP Conf. Series: Journal of Physics: Conf. Series **805** (2017) 012004 doi:10.1088/1742-6596/805/1/012004

Multi-Reggeon exchanges at high p_t from gluon saturation

Jamal Jalilian-Marian

Natural Sciences Department, Baruch College, 17 Lexington Ave, New York, NY 10010
CUNY Graduate Center, 365 5th Ave., New York, NY 10016

E-mail: jamal.jalilian-marian@baruch.cuny.edu

Abstract. We show that the JIMWLK equation for rapidity evolution of $2n$ Wilson lines at high p_t reduces to the BJKP equation for the rapidity evolution of t-channel exchange of $2n$ Reggeized gluons.

1. Introduction

Understanding multi-particle production cross sections in QCD at high energy has been an active and exciting field of research. Approaches based on gluon saturation, commonly referred to as the Color Glass Condensate (CGC) formalism [1], have been quite successful in understanding qualitative and semi-quantitative features of the recent data from the Relativistic Heavy Ion Collider (RHIC) and the Large Hadron Collider (LHC) [2, 3, 4]. The most essential feature of gluon saturation is the emergence of a semi-hard scale, called saturation scale [5] which grows with energy (or equivalently with x or rapidity) and allows a first principle and weak coupling computation of production of low momentum partons which would not be possible in the standard pQCD due to infrared divergences. In this formalism one treats the large x partons as sources of color charge ρ^a and then solves the Yang-Mills classical equations of motion in the presence of this color charge. This solution is then taken as a background field in the presence of which one does quantum (loop) corrections. In the high energy limit the most important effect of loop corrections is appearance of a large logarithm of x (or energy) which needs to be re-summed, analogous to appearance of large logarithms of virtuality Q^2 which are re-summed in the standard pQCD by the DGLAP evolution equation. The resulting (functional) equation is known as the JIMWLK equation [6] and can be used to derive explicit evolution equations for the rapidity (x or energy) dependence of any correlation function of the theory.

There is an alternative approach to scattering in high energy QCD based on exchanges of Reggeized gluons. Here the effective degrees of freedom exchanged in an interaction are Reggeized gluons which are built from ordinary gluons of pQCD via inclusion of small x effects. For the case that the number of exchanged reggeized gluons is fixed the evolution of the state is governed by the BJKP equation [7]. For the special case of exchange of two Reggeized gluons the BJKP equation reduces to the well-known BFKL equation [8]. Since both the CGC formalism and the multi-Reggeized gluon exchange approach apply to high energy processes in QCD it is an important and interesting question as whether the two approaches are equivalent approaches, in the kinematic region where they are based on the same approximations. Here we study the high p_t limit of the JIMWLK equation and show that the JIMWLK equation reduces to the BJKP equation in the high p_t limit, i.e., in the dilute gluon regime when one can expand the Wilson lines in powers of the gluon field $g\,A_\mu$ (or alternatively in powers of the color charge density ρ).

2. The JIMWLK evolution equation

The JIMWLK equation describes evolution of any correlation function O with rapidity Y and is given by

$$\frac{\partial < O >_Y}{\partial Y} = < H\,O >_Y \tag{1}$$

where the Hamiltonian H is

$$H = -\frac{1}{16\pi^3}\int d^2x\, d^2y\, d^2z M_{xyz}\left(1 + U_x^\dagger U_y - U_x^\dagger U_z - U_z^\dagger U_y\right)^{ab}\frac{\delta}{\delta\alpha_x^a}\frac{\delta}{\delta\alpha_y^b}, \tag{2}$$

where x, y, z are two-dimensional vectors on the transverse plane, a and b are color indices and the dipole kernel M_{xyz} is defined as

$$M_{xyz} \equiv \frac{(x-y)^2}{(x-z)^2(z-y)^2} \tag{3}$$

where U is a Wilson line in the adjoint representation

$$U(x) \equiv e^{-ig\,\alpha^a(x)\,T^a}, \tag{4}$$

with T^a as generators of $SU(N)$ in the adjoint representation. The gluon field $\alpha^a(x_t)$ satisfies the classical equations of motion

$$\partial_\perp^2 \alpha^a(x_t) = -g\rho^a(x_t). \tag{5}$$

We start by considering the JIMWLK evolution equation for the trace of $2n$ Wilson lines in the fundamental representation, denoted by V (with T^a in (4) replaced by t^a which is in the fundamental representation)

$$\hat{S}^{(2n)}_{(\prod_{k=1}^{2n} x_k)} = \frac{1}{N_c} Tr\left(V_{x_1} V_{x_2}^\dagger V_{x_3} V_{x_4}^\dagger \cdots V_{x_{2n-1}} V_{x_{2n}}^\dagger\right) \tag{6}$$

and apply the functional derivatives on the Wilson lines as

$$\begin{aligned} \frac{\delta}{\delta\alpha^a(x)} V_{x_i}^\dagger &= ig\,\delta^2(x_i - x)\,t^a\,V_{x_i}^\dagger \\ \frac{\delta}{\delta\alpha^a(x)} V_{x_i} &= -\,ig\,\delta^2(x_i - x)\,V_{x_i}\,t^a \end{aligned} \tag{7}$$

and use the following identities relating fundamental and adjoint Wlison lines repeatedly

$$\begin{aligned} (U_x^\dagger)^{ac}\,t^a &= V_x^\dagger\,t^c\,V_x \\ (U_y)^{cb}\,t^b &= V_y^\dagger\,t^c\,V_y \end{aligned} \tag{8}$$

The result is long and can be written symbolically as

$$\begin{aligned} H\,\hat{S}^{(2n)} = \frac{\bar{\alpha}}{4\pi}\int_z \quad & \Big(H_1\,\hat{S}^{(2n)} + H_2\,\hat{S}^{(2n)} \\ & - H_3\,\hat{S}^{(2n)} - H_4\,\hat{S}^{(2n)}\Big), \end{aligned} \tag{9}$$

where the individual terms are given in [9].

Here we are interested in establishing the equivalence between the JIMWLK and BJKP equations in the high p_t limit. This means we have to expand the Wilson lines in powers of the gauge field

$$\begin{aligned} V_{x_i} &= 1 - ig\alpha_{x_i} + \cdots \\ V^\dagger_{x_i} &= 1 + ig\alpha_{x_i} + \cdots \end{aligned} \tag{10}$$

and keep the first non-trivial term in each expansion. Furthermore we disregard the non-linear terms on the right hand side of the evolution equation and rewrite it using T rather than S matrices. We get

$$\hat{T}^{(2n)}_{\left(\prod_{k=1}^{2n} x_k\right)} \simeq g^{2n} \frac{1}{N_c} Tr[\alpha_{x_1}\alpha_{x_2}\cdots\alpha_{x_{2n-1}}\alpha_{x_{2n}}]. \tag{11}$$

In addition, one needs to include terms which are quadratic in $\alpha(z_t)$. We then Fourier transform the coordinate space expression to momentum space and define the operator $T^{(2n)}$ in terms of the color charge ρ related to the classical gluon field by $\alpha(p_t) \sim \frac{\rho(p_t)}{p_t^2}$,

$$T^{(2n)}_{\left(\prod_{k=1}^{2n} l_k\right)} = \frac{1}{N_c} Tr\left[\prod_{k=1}^{2n} \rho(l_k)\right] \tag{12}$$

The evolution equation written for $T^{(2n)}$ reads

$$\begin{aligned} \frac{d}{dY} T^{(2n)}_{\left(\prod_{k=1}^{2n} l_k\right)} &= -\frac{\bar{\alpha}}{2\pi} \sum_{j=1}^{2n} \int d^2p_t \left[\frac{l_j^2}{p_t^2\,[p_t^2 + (p_t - l_j)^2]}\right] T^{(2n)}_{\left(\prod_{k=1}^{2n} l_k\right)} \\ &+ \frac{\bar{\alpha}}{4\pi} \int d^2p_t \left[\frac{l_1^2}{p_t^2\,(p_t + l_1)^2} + \frac{l_{2n}^2}{p_t^2\,(p_t - l_{2n})^2} - \frac{(l_1 + l_{2n})^2}{(p_t + l_1)^2\,(p_t - l_{2n})^2}\right] T^{(2n)}_{(l_1+p_t)\left(\prod_{k=2}^{2n-1} l_k\right)(l_{2n}-p_t)} \\ &+ \frac{\bar{\alpha}}{4\pi} \sum_{j=2}^{2n} \int d^2p_t \left[\frac{l_{j-1}^2}{p_t^2\,(p_t + l_{j-1})^2} + \frac{l_j^2}{p_t^2\,(p_t - l_j)^2} - \frac{(l_{j-1} + l_j)^2}{(p_t + l_{j-1})^2\,(p_t - l_j)^2}\right] \\ & T^{(2n)}_{\left(\prod_{k=1}^{j-2} l_k\right)(l_{j-1}+p_t)(l_j-p_t)\left(\prod_{k=j+1}^{2n} l_k\right)}. \end{aligned} \tag{13}$$

This is our final result and is identical to the BJKP evolution equation. It is obtained by first writing the JIMWLK evolution equation for $2n$ Wilson lines in the fundamental representation, and then expanding each of the Wilson lines in powers of the gluon field and disregarding all terms which are quadratic and higher in powers of gluon fields which are at the same external coordinates. This establishes a formal equivalence between the JIMWLK evolution equation for $2n$ Wilson lines at high p_t and the BJKP equation which describes the energy dependence of a state of $2n$-Reggeized gluon exchanged in the t channel. In principle the JIMWLK equation also includes Reggeized gluon number changing exchanges [10] which are not included in the BJKP equation. These become important at low transverse momentum where gluon saturation effects are dominant. These effects are contained in triple (and beyond) pomeron vertices in Reggeized-gluon exchange approach.

Acknowledgments

The author acknowledges support by the DOE Office of Nuclear Physics through Grant No. DE-FG02-09ER41620 and from The City University of New York through the PSC-CUNY Research Award Program, grant 67732-0045.

[1] E. Iancu and R. Venugopalan, In *Hwa, R.C. (ed.) et al.: Quark gluon plasma* 249-3363; H. Weigert, Prog. Part. Nucl. Phys. **55**, 461 (2005); J. Jalilian-Marian, Y. V. Kovchegov, Prog. Part. Nucl. Phys. **56**, 104-231 (2006); F. Gelis, E. Iancu, J. Jalilian-Marian, R. Venugopalan, Ann. Rev. Nucl. Part. Sci. **60**, 463-489 (2010); J. Jalilian-Marian, Y. V. Kovchegov, Prog. Part. Nucl. Phys. **56**, 104-231 (2006); J. L. Albacete and C. Marquet, Prog. Part. Nucl. Phys. **76**, 1 (2014).

[2] A. Dumitru and J. Jalilian-Marian, Phys. Rev. Lett. **89**, 022301 (2002); D. Kharzeev, Y. V. Kovchegov and K. Tuchin, Phys. Rev. D **68**, 094013 (2003); J. L. Albacete, N. Armesto, A. Kovner, C. A. Salgado and U. A. Wiedemann, Phys. Rev. Lett. **92**, 082001 (2004); J. Jalilian-Marian, J. Phys. G **30**, S751 (2004), Nucl. Phys. A **748**, 664 (2005), Phys. Rev. C **70**, 027902 (2004), Nucl. Phys. A **739**, 319 (2004); A. Dumitru, A. Hayashigaki and J. Jalilian-Marian, Nucl. Phys. A **765**, 464 (2006), Nucl. Phys. A **770**, 57 (2006); F. Gelis and J. Jalilian-Marian, Phys. Rev. D **76**, 074015 (2007); D. Boer, A. Utermann and E. Wessels, Phys. Rev. D **77**, 054014 (2008); E. R. Cazaroto, V. P. Gonçalves, F. S. Navarra, Nucl. Phys. A **872**, 196 (2011); J. Jalilian-Marian and A. H. Rezaeian, Phys. Rev. D **85**, 014017 (2012), Phys. Rev. D **86**, 034016 (2012); A. H. Rezaeian, Phys. Rev. D **86**, 094016 (2012); J. L. Albacete, A. Dumitru, H. Fujii and Y. Nara, Nucl. Phys. A **897**, 1 (2013); T. Altinoluk and A. Kovner, Phys. Rev. D **83**, 105004 (2011); G. A. Chirilli, B. -W. Xiao and F. Yuan, Phys. Rev. Lett. **108**, 122301 (2012), Phys. Rev. D **86**, 054005 (2012); Z. -B. Kang, I. Vitev and H. Xing, arXiv:1403.5221 [hep-ph].

[3] J. Jalilian-Marian and Y. V. Kovchegov, Phys. Rev. D **70**, 114017 (2004) [Erratum-ibid. D **71**, 079901 (2005)]; A. Kovner and M. Lublinsky, JHEP **0611**, 083 (2006), Phys. Rev. D **84**, 094011 (2011), Phys. Rev. D **83**, 034017 (2011); J. Jalilian-Marian, Nucl. Phys. A **770**, 210 (2006), Eur. Phys. J. C **61**, 789 (2009); C. Marquet, Nucl. Phys. A **796**, 41 (2007); J. L. Albacete and C. Marquet, Phys. Rev. Lett. **105**, 162301 (2010); F. Dominguez, C. Marquet, B. -W. Xiao and F. Yuan, Phys. Rev. D **83**, 105005 (2011); A. Stasto, B. -W. Xiao and F. Yuan, Phys. Lett. B **716**, 430 (2012); F. Dominguez, C. Marquet, A. M. Stasto and B. -W. Xiao, Phys. Rev. D **87**, no. 3, 034007 (2013); T. Lappi and H. Mantysaari, Nucl. Phys. A **908**, 51 (2013); E. Iancu and D. N. Triantafyllopoulos, JHEP **1311**, 067 (2013); A. Ayala, J. Jalilian-Marian, J. Magnin, A. Ortiz, G. Paic and M. E. Tejeda-Yeomans, Phys. Rev. Lett. **104**, 042301 (2010).

[4] Y. V. Kovchegov, J. Kuokkanen, K. Rummukainen and H. Weigert, Nucl. Phys. A **823**, 47 (2009); C. Marquet and H. Weigert, Nucl. Phys. A **843**, 68 (2010); A. Dumitru and J. Jalilian-Marian, Phys. Rev. D **82**, 074023 (2010), Phys. Rev. D **81**, 094015 (2010); A. Dumitru, J. Jalilian-Marian, T. Lappi, B. Schenke and R. Venugopalan, Phys. Lett. B **706**, 219 (2011); E. Iancu and D. N. Triantafyllopoulos, JHEP **1111**, 105 (2011).

[5] L. D. McLerran and R. Venugopalan, Phys. Rev. D **49**, 2233 (1994), Phys. Rev. D **49**, 3352 (1994); J. Jalilian-Marian, A. Kovner, L. D. McLerran and H. Weigert, Phys. Rev. D **55**, 5414 (1997); S. Jeon and R. Venugopalan, Phys. Rev. D **71**, 125003 (2005); A. Dumitru, J. Jalilian-Marian, E. Petreska, Phys. Rev. **D84**, 014018 (2011); J. Jalilian-Marian, S. Jeon and R. Venugopalan, Phys. Rev. D **63**, 036004 (2001).

[6] J. Jalilian-Marian, A. Kovner, L. D. McLerran and H. Weigert, Phys. Rev. D **55**, 5414 (1997); J. Jalilian-Marian, A. Kovner, A. Leonidov and H. Weigert, Nucl. Phys. B **504**, 415 (1997), Phys. Rev. D **59**, 014014 (1999), Phys. Rev. D **59**, 014015 (1999), Phys. Rev. D **59**, 034007 (1999), A. Kovner, J. G. Milhano and H. Weigert, Phys. Rev. D **62**, 114005 (2000); A. Kovner and J. G. Milhano, Phys. Rev. D **61**, 014012 (2000); E. Iancu, A. Leonidov and L. D. McLerran, Nucl. Phys. A **692**, 583 (2001), Phys. Lett. B **510**, 133 (2001); E. Ferreiro, E. Iancu, A. Leonidov and L. McLerran, Nucl. Phys. A **703**, 489 (2002); H. Weigert, Nucl. Phys. **A703**, 823-860 (2002); J. -P. Blaizot, E. Iancu, H. Weigert, Nucl. Phys. **A713**, 441-469 (2003).

[7] J. Bartels, Nucl. Phys. B **175**, 365 (1980); T. Jaroszewicz, Acta Phys. Polon. B **11**, 965 (1980); J. Kwiecinski and M. Praszalowicz, Phys. Lett. B **94**, 413 (1980); Z. Chen, A. H. Mueller, Nucl. Phys. **B451**, 579-604 (1995).

[8] L. N. Lipatov, Sov. J. Nucl. Phys. **23**, 338 (1976) [Yad. Fiz. **23**, 642 (1976)]; E. A. Kuraev, L. N. Lipatov and V. S. Fadin, Sov. Phys. JETP **45**, 199 (1977) [Zh. Eksp. Teor. Fiz. **72**, 377 (1977)]; I. I. Balitsky and L. N. Lipatov, Sov. J. Nucl. Phys. **28**, 822 (1978) [Yad. Fiz. **28**, 1597 (1978)].

[9] A. Ayala, E. R. Cazaroto, L. A. Hernndez, J. Jalilian-Marian and M. E. Tejeda-Yeomans, Phys. Rev. D **90**, 074037 (2014).

[10] F. Dominguez, A. H. Mueller, S. Munier and B. -W. Xiao, Phys. Lett. B **705**, 106 (2011); J. Jalilian-Marian, Phys. Rev. D **85**, 014037 (2012).

IOP Conf. Series: Journal of Physics: Conf. Series **805** (2017) 012005 doi:10.1088/1742-6596/805/1/012005

Is q̂ a physical quantity or just a parameter? and other unanswered questions in high-p_T physics

M. J. Tannenbaum[1]
Physics Department, Brookhaven National Laboratory, Upton, NY 11973-5000 USA
E-mail: mjt@bnl.gov

Abstract. The many different theoretical studies of energy loss of a quark or gluon traversing a medium have one thing in common: the transport coefficient of a gluon in the medium, denoted $\hat{q}$, which is defined as the mean 4-momentum transfer-square, q^2, by a gluon to the medium per gluon mean free path, $\lambda_{\rm mfp}$. In the original BDMPSZ formalism, the energy loss of an outgoing parton, $-dE/dx$, per unit length (x) of a medium with total length L, due to coherent gluon bremsstrahlung is proportional to the q^2 and takes the form:

$$\frac{-dE}{dx} \simeq \alpha_s \langle q^2(L)\rangle = \alpha_s\, \mu^2\, L/\lambda_{\rm mfp} = \alpha_s\, \hat{q}\, L \qquad , \qquad (1)$$

where μ, is the mean momentum transfer per collision. Thus, the total energy loss in the medium goes like L^2.

Additionally, the accumulated momentum-square, $\langle k_T^2\rangle$, transverse to a gluon traversing a length L in the medium is well approximated by $\langle k_T^2\rangle \approx \langle q^2(L)\rangle = \hat{q}\, L$. A simple estimate shows that the $\langle k_T^2\rangle \approx \hat{q}\, L$ should be observable at RHIC at $\sqrt{s_{NN}}$ =200 GeV via the broadening of di-hadron azimuthal correlations resulting in an azimuthal width $\sim \sqrt{2}$ larger in Au+Au than in p+p collisions . Measurements relevant to this issue will be discussed as well as recent STAR jet results presented at QM2014 [1].

Other topics to be discussed include the danger of using forward energy to define centrality in $p(d)$+A collisions for high p_T measurements, the danger of not using comparison $p+p$ data at the same $\sqrt{s}$ in the same detector for $R_{\rm AA}$ or lately for $R_{\rm pA}$ measurements. Also, based on a comment at last year's 9th workshop that the parton energy loss is proportional to $dN_{\rm ch}/d\eta$ [2], new results on the dependence of the shift in the p_T spectra in A+A collisions from the $T_{\rm AA}$-scaled $p+p$ spectrum (to be discussed in detail in another presentation [3]) will be shown.

1. Introduction–BDMPSZ, the first QCD based Jet Quenching Model

I don't want to discuss models in detail, since they are nothing like QED or QCD—theories that you can set your watch by (at least QED). I concentrate on one example, the first QCD based model [4] which stimulated the use of hard-probes at RHIC as a signature of the QGP.

It is important to note that the original STAR Letter of Intent (LBL-29651) in 1990, following Wang and Gyulassy (LBL-29390), did cite as one objective: "the use of hard scattering of partons as a probe of high density nuclear matter... Passage through hadronic or nuclear matter is predicted to result in an attenuation of the jet energy and broadening of jets. Relative to this damped case, a QGP is transparent and an enhanced yield is expected."

[1] Supported by the U.S. Department of Energy, Contracts DE-AC02-98CH10886 and DE-SC0012704.

Of course this is precisely the opposite of what was actually discovered at RHIC. Furthermore, what had been observed in A+A and p+A collisions was an enhancement of the hard scattering, a.k.a. the Cronin Effect [5], rather than an attenuation. Thus, until the appearance of the fully QCD based models, starting with BDMPS [6], I described the original Plümer-Gyulassy-Wang [7, 8] Jet Quenching as "the vanishing of something that doesn't exist in the first place", namely the attenuation of hard-scattering in dense but confined nuclear matter (CNM).

In the early c. 1990 publications [7, 9] the QGP effect was thought to be "a sudden *decrease* of dE/dx near the quark-gluon plasma phase transition" which could *reduce* the CNM Jet Quenching ("unquench the jets" [9]) and thus be a possible signature of the QGP. This idea was downplayed between the original STAR letter of intent in September 1990 and the update in July 1991 as reflected in the new goal for Parton Physics: "For example, it has been suggested that there will be observable changes in the energy loss of propagating partons as the energy density of the medium increases, particularly if the medium passes through a phase transition to the QGP [10]".

The reason for the downplaying of jet quenching as a possible probe of the QGP by STAR in 1991 was the discovery [10] that instead of small dE/dx in the QGP it was "recently found that at least deep in the QGP phase, the induced radiative energy loss could be quite large" [10]. Subsequent work found that the radiation was suppressed by the LPM effect [11] which led to a series of developments, nicely reviewed in Ref. [4], that eventually led to the BDMPSZ QCD based model [4].

1.1. The real QGP jet quenching and the importance of attending conferences

I first heard about the original CNM Jet Quenching [7] at an excellent meeting in Strasbourg in October 1990 [12] to discuss "Quark-Gluon Plasma Signatures" in a talk by Michael Plümer that was greeted with disbelief by the many CERN-ISR veterans who had puzzled over the Cronin effect for many years. This led to my description noted a few paragraphs above. Meanwhile, the RHIC experiments and ALICE at the LHC [13] were designed with a focus on J/Ψ suppression [14] as the gold-plated signature for deconfinement and the QGP.

In 1998 at the QCD workshop in Paris [15], I found what I thought was a cleaner signal of the QGP when Rolf Baier asked me whether jets could be measured in Au+Au collisions because he had made studies in pQCD [6] of the energy loss of partons, produced by hard-scattering "with their color charge fully exposed", in traversing a medium "with a large density of similarly exposed color charges". The conclusion was that "Numerical estimates of the loss suggest that it may be significantly greater in hot matter than in cold. *This makes the magnitude of the radiative energy loss a remarkable signal for QGP formation*" [4]. In addition to being a probe of the QGP, the fully exposed color charges allow the study of parton-scattering with $Q^2 \ll 1-5$ $(\mathrm{GeV}/c)^2$ in the medium where new collective QCD effects may possibly be observed.

Because the expected energy in a typical jet cone $R = \sqrt{(\Delta\eta)^2 + (\Delta\phi)^2}$ in central Au+Au collisions at $\sqrt{s_{NN}}$ =200 GeV would be $\pi R^2 \times 1/2\pi \times d\mathrm{E}_T/d\eta = R^2/2 \times d\mathrm{E}_T/d\eta \sim 300$ GeV for $R = 1$, where the kinematic limit is 100 GeV, I said (and wrote [15][2]) that jets can not be reconstructed in Au+Au central collisions at RHIC—still correct after 16 years. On the other hand, hard-scattering was discovered in $p + p$ collisions at the CERN-ISR in 1972 with single particle and two-particle correlations, while jets had a long learning curve from 1977–1982 with a notorious false claim (e.g. see Refs. [17, 18]), so I said (and wrote [15]) that we should use single and two-particle measurements at RHIC—which we did and it WORKED! The present solution for jets in A+A collisions (LHC 2010 and RHIC c.2014) is to take smaller cones, with 100 GeV in $R = 0.58$, 48 GeV in $R = 0.4$, 27 GeV in $R = 0.3$, 12 GeV in $R = 0.2$ at RHIC.

[2] It was an excellent guess because the measured $d\mathrm{E}_T/d\eta = 606 \pm 32$ GeV in central Au+Au at $\sqrt{s_{NN}}$ =200 GeV [16].

1.2. $\hat{q}$ or di-jet broadening and gluon radiation

There are many different theoretical studies of energy loss of a quark or gluon with their color charges fully exposed passing through a medium with a large density of similarly exposed color charges (i.e. a QGP). The approaches are different, but the one thing that they have in common [19] is the transport coefficient of a gluon in the medium, denoted $\hat{q}$, which is defined as the mean 4-momentum transfer-square, q^2, by a gluon to the medium per gluon mean free path, $\lambda_{\rm mfp}$. Thus the mean 4-momentum transfer-square for a gluon traversing length L in the medium is, $\langle q^2(L)\rangle = \hat{q}\,L = \mu^2\,L/\lambda_{\rm mfp}$, where μ, the mean momentum transfer per collision, is "conveniently taken" [4] as the Debye screening mass acquired by gluons in the medium. In this, the original BDMPSZ formalism [4], the energy loss of an outgoing parton, $-dE/dx$, per unit length (x) of a medium with total length L, due to coherent gluon bremsstrahlung is proportional to the 4-momentum-square transferred to the medium and takes the form:

$$\frac{-dE}{dx} \simeq \alpha_s\langle q^2(L)\rangle = \alpha_s\,\hat{q}\,L = \alpha_s\,\mu^2\,L/\lambda_{\rm mfp} \qquad , \qquad (2)$$

so that the total energy loss in the medium goes like L^2 [6].

Additionally the accumulated transverse momentum-square, $\langle k_T^2\rangle$, for a gluon traversing a length L in the medium is well approximated by $\langle k_T^2\rangle \approx \langle q^2(L)\rangle = \hat{q}\,L$. This leads to a remarkable relationship [4] between the energy loss and di-jet broadening ("acoplanarity [7]"):

$$\frac{-dE}{dx} \simeq \alpha_s\langle k_T^2\rangle \qquad , \qquad (3)$$

which is thought to be independent of the dynamics of the individual scatterings in pQCD and thus should be expected to hold equally in a finite length QGP and CNM [20]. A simple estimate shows that the $\langle k_T^2\rangle \approx \hat{q}\,L$ should be observable at RHIC via the broadening of di-hadron azimuthal correlations. Assume that for a trigger particle with p_{T_t} the away-parton traverses slightly more than half the 14 fm diameter medium for central collisions of Au+Au, say 8 fm. With a $\hat{q} = 1$ GeV^2/fm [19], this would correspond to $\langle k_T^2\rangle = \hat{q}\,L = 8$ (GeV/c)^2, compared to the measured [21] $\langle k_T^2\rangle = 8.0 \pm 0.2$ (GeV/c)^2 for di-hadrons in $p+p$ collisions[3] with roughly the same p_{T_t} and $p_T^{\rm assoc}$. This should be visible as a width of the $p_T^{\rm assoc}$ azimuthal distribution $\sim \sqrt{2}$ larger in Au+Au than in $p+p$ collisions at $\sqrt{s_{NN}}$ =200 GeV.

However, there is no direct evidence as yet for broadening of di-hadron or di-jet correlations from the effect of $\hat{q}$ in either d+Au [22] or Au+Au collisions at RHIC, where the principal difficulty in Au+Au stems from the systematic uncertanties due to the collective flow background of the medium, $v_2, v_3 \ldots v_n$ for di-hadron measurements; nor at LHC, where the very large jet $p_T \simeq 100$ GeV/c, for di-jet measurements, may have obscured this signal.

2. Discovery of the real QGP jet quenching, RHIC's main claim to fame.

The discovery at RHIC [23] that π^0's produced at large transverse momenta are suppressed in central Au+Au collisions by a factor of ~ 5 compared to pointlike scaling from p+p collisions is arguably *the* major discovery in Relativistic Heavy Ion Physics. For π^0 (Fig. 1a) [24] the hard-scattering in p+p collisions is indicated by the power law behavior p_T^{-n} for the invariant cross section, $Ed^3\sigma/dp^3$, with $n = 8.1 \pm 0.1$ for $p_T \geq 3$ GeV/c. The Au+Au data at a given p_T can be characterized either as shifted lower in p_T by $\delta p'_T$ from the pointlike scaled p+p data at $p'_T = p_T + \delta p'_T$, or shifted down in magnitude, i.e. suppressed. In Fig. 1b, the suppression of the many identified particles measured by PHENIX at RHIC is presented as the Nuclear Modification Factor, $R_{\rm AA}(p_T)$, the ratio of the yield of e.g. π per central Au+Au collision

[3] In both cases the azimuthal projection is only half the $\langle k_T^2\rangle$ in $p+p$ or from $\hat{q}$.

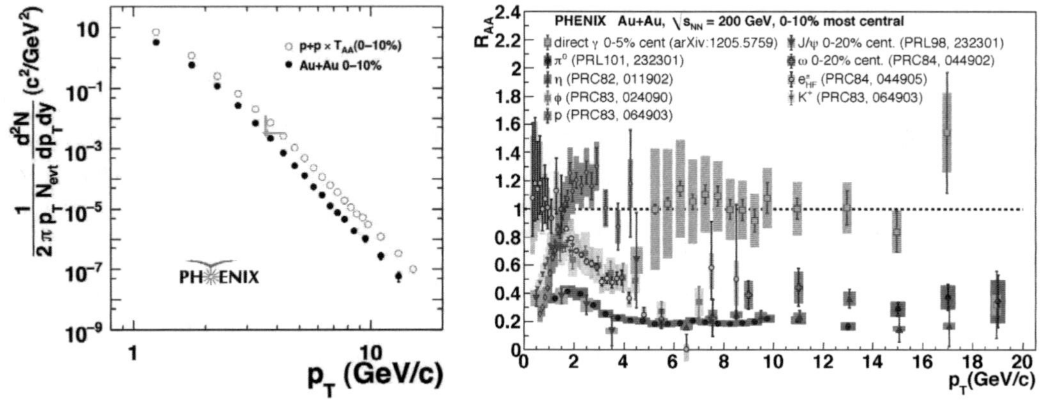

Figure 1. a) (left) Log-log plot of invariant yield of π^0 at $\sqrt{s_{NN}} = 200$ GeV as a function of transverse momentum p_T in p+p collisions, multiplied by $\langle T_{\rm AA}\rangle$ for Au+Au central (0–10%) collisions, compared to the Au+Au measurement [24]. Vertical arrow is for $R_{\rm AA}(p_T)$, horizontal arrow for $\delta p'_T$. b) (right) $R_{\rm AA}(p_T)$ for all identified particles so far measured by PHENIX in Au+Au central collisions at $\sqrt{s_{NN}} = 200$ GeV.

(upper 10%-ile of observed multiplicity) to the pointlike-scaled p+p cross section at the same p_T, where $\langle T_{\rm AA}\rangle$ is the average overlap integral of the nuclear thickness functions:

$$R_{\rm AA}(p_T) = \frac{(1/N_{\rm AA})\; d^2N^{\pi}_{\rm AA}/dp_T dy}{\langle T_{\rm AA}\rangle\; d^2\sigma^{\pi}_{\rm pp}/dp_T dy} \quad . \tag{4}$$

The striking differences and similarities of $R_{\rm AA}(p_T)$ in central Au+Au collisions for the many particles measured by PHENIX (Fig. 1b) illustrate the importance of particle identification for understanding the physics of the medium produced at RHIC. Notable are that ALL particles are suppressed for $p_T > 4$ GeV/c (except for direct-γ which are not coupled to color), even electrons from c and b quark decay; with one notable exception: the protons are enhanced for $2 \leq p_T \leq 4$ GeV/c, called the baryon anomaly, although recently the same Cronin-like effect has been seen in d+Au collisions [25].

2.1. $\delta \mathbf{p'_T}/\mathbf{p'_T}$, *the fractional shift in the* $\mathbf{p'_T}$ *spectrum*

After more than a decade of using the ratio $R_{\rm AA}$, we are now paying more attention to $\delta p'_T/p'_T$, the fractional shift of the p'_T spectrum, as an indicator of energy loss in the QGP Fig. 2 [3]. For a constant fractional energy loss, which is true at RHIC in the range $6 < p_T < 12$ GeV/c (as shown in Fig. 1a where the p+p reference and Au+Au measurement are parallel on a log-log plot) there is a simple relationship between $R_{\rm AA}$, $\delta p'_T/p'_T$ and n, the power in the invariant p_T spectra:

$$R_{\rm AA}(p'_T) = R_{\rm AA}(p_T) = (1 - \delta p'_T/p'_T)^{n-2} \quad . \tag{5}$$

Using $\delta p'_T/p'_T$ is important for comparison to the LHC measurements where the power is $n \approx 6$ compared to $n = 8.1$ at RHIC, so that the same $R_{\rm AA}$ does not mean the same $\delta p'_T/p'_T$. Strictly $\delta p'_T/p'_T$ is not a measure of the parton energy loss in the QGP but is used as a proxy. Figure 2a shows that $\delta p'_T/p'_T$ at $p'_T = 7$ GeV/c for RHIC and LHC both increase monotonically with centrality ($N_{\rm part}$) but is a factor of 2 to 1.4 larger at LHC, depending on centrality, a likely indication of a hotter and/or denser medium. Figure 2b attempts to determine whether $\delta p'_T/p'_T$ is a universal function of the charged particle density, $dN_{\rm ch}/d\eta$ at both RHIC and LHC, as suggested by Edward Shuryak at this meeting last year [2]. The dependence is not quite

IOP Conf. Series: Journal of Physics: Conf. Series **805** (2017) 012005 doi:10.1088/1742-6596/805/1/012005

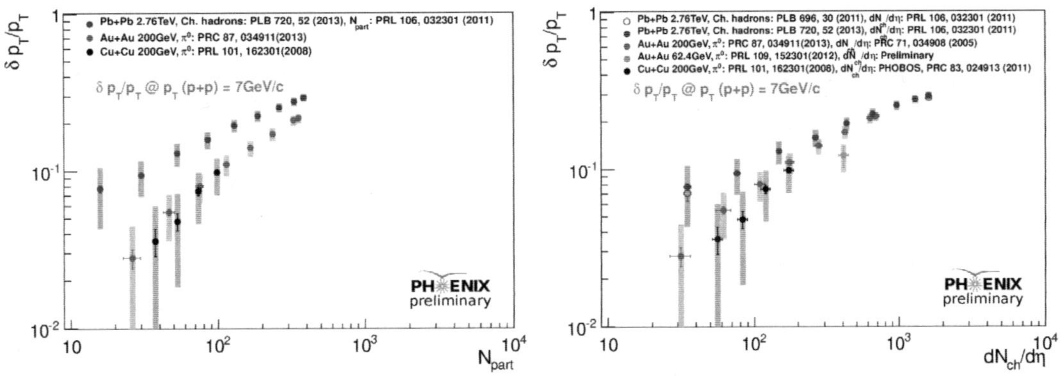

Figure 2. Plots from PHENIX [3] of $\delta p'_T/p'_T$ at $p'_T \equiv p_T(p+p) = 7$ GeV for π^0 (RHIC) and charged hadrons (LHC): a) as a function of centrality ($N_{\rm part}$), b) as a function of $dN_{\rm ch}/d\eta$.

universal. A fit of $\delta p_T/p_T \propto (dN_{\rm ch}/d\eta)^{\alpha}$ gives $\alpha \approx 0.35$ at LHC and 0.55 at RHIC, although the data at $\sqrt{s_{NN}} = 200$ GeV and 2.76 TeV do appear to merge for $(dN_{\rm ch}/d\eta) \geq 300$. Hopefully, measurements of $\delta p_T/p_T$ will eventually lead to the determination of dE/dx of partons in the QGP.

3. STAR jet and jet-hadron measurements

3.1. Jet-hadron correlations as a proxy for di-jet broadening

Admittedly, measuring jets at RHIC at $\sqrt{s_{NN}}$ =200 GeV is much harder than at LHC at $\sqrt{s_{NN}}$ =2.76 TeV: the cross section in the relevant region is $\gtrsim$100 times larger at LHC while the soft physics background is only a factor of 2 larger [26]. Nevertheless, the principal difficulty in observing the broadening of di-jet or di-hadron azimuthal correlations by the transport coefficient $\hat{q}$ of the QGP stems from artifacts with names such as "Mach Cone", "Ridge", "Head and Shoulders" which are now known to be due to the modulation of the soft physics background by collective flow with both even and odd harmonics [27]. Of course, understanding that the extra "bumps" in the correlation function are due to odd harmonics still requires one to know the values of these harmonics in order to subtract them. This is still the largest systematic uncertainty in attempts to observe the $\hat{q}$-broadening, for instance, the most recent attempt by STAR using jet-hadron correlations [28] (Fig. 3). When the full systematic uncertainties, including those

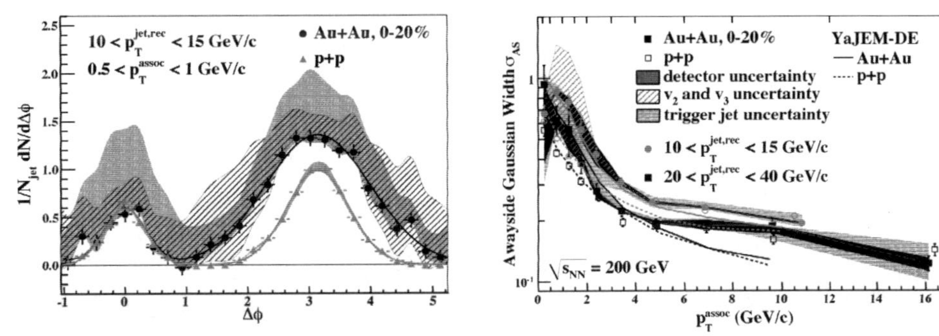

Figure 3. a) (left) Azimuthal correlation ($1/N_{\rm jet}dN/d\Delta\phi$) in Au+Au and $p+p$ with systematic uncertainties shown [28]. b)(right) Awayside rms width, $\sigma_{\rm AS}$, as a function of $p_T^{\rm assoc}$ [28].

on v_2 and v_3 (Fig. 3a), are taken into account, the result for the medium induced broadening of the away-side widths, $\sigma_{\rm AS}$, in Au+Au relative to $p+p$ (Fig. 3b) which looked significant in the preliminary results, as shown last year [29], become only "suggestive of medium-induced broadening [28]" in the final result because "they are highly dependent on the shape of the subtracted background [28]", notably the v_2 and v_3 of the trigger jets.

3.2. At last: jet measurements in Au+Au at RHIC in 2014?

Some interesting new jet measurements in Au+Au collisions at RHIC were presented at Quark Matter 2014 in a plenary review talk on jets by Yen-Jie Lee [1] who works on CMS. Figure 4 shows that the STAR charged jets in a cone with $R = 0.2$ have much less suppression ($R_{\rm AA} \gg 0.3$) than π^0 ($0.2 \leq R_{\rm AA} \leq 0.3$) in the range $10 < p_T < 20$ GeV.

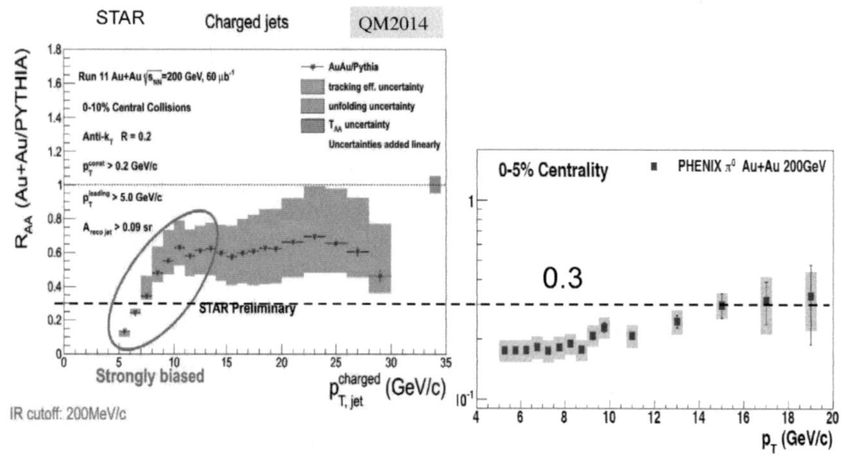

Figure 4. a) (left) STAR $R_{\rm AA}$ for charged jets at $\sqrt{s_{NN}}$ =200 GeV in central Au+Au collisions (see details in legend) compared to b) $R_{\rm AA}$ for PHENIX π^0. The dashed line at 0.3 is the maximum $R_{\rm AA}$ for π^0 in this p_T range.

This is quite different from jets at the LHC (Fig. 5) which have comparable or smaller $R_{\rm AA}$ than charged particles from jet fragmentation in the range $30 < p_T < 100$ GeV. Note that the γ, W and Z^0 bosons in Fig. 5b which are not coupled to color are not suppressed.

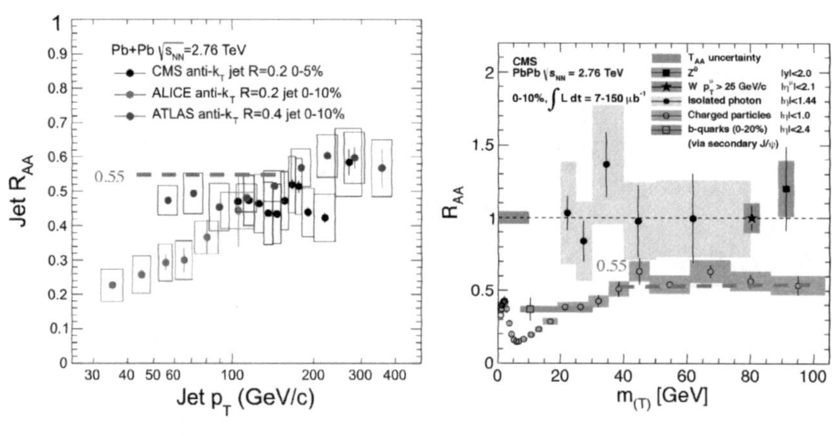

Figure 5. a) (left) $R_{\rm AA}$ for jets at $\sqrt{s_{NN}}$ =2.76 GeV by CMS and ALICE compared to b) CMS $R_{\rm AA}$ for charged hadrons ($R_{\rm AA} \approx 0.55$), b-quarks and 3 favorite Electro-Weak Bosons [1].

For STAR, the disagreement of the jet and single particle R_{AA} gets worse as the jet cone is increased from R=0.2 to 0.3 to 0.4 (Fig. 6). Some people would say that this is great because all the jet fragments and/or any energy lost in the QGP by the originating parton have been captured in the R=0.4 cone. Skeptics like myself can hardly wait to see what happens when the jet cone is further increased. After 14 runs at RHIC, the jet learning curve in Au+Au central collisions still has a way to go.

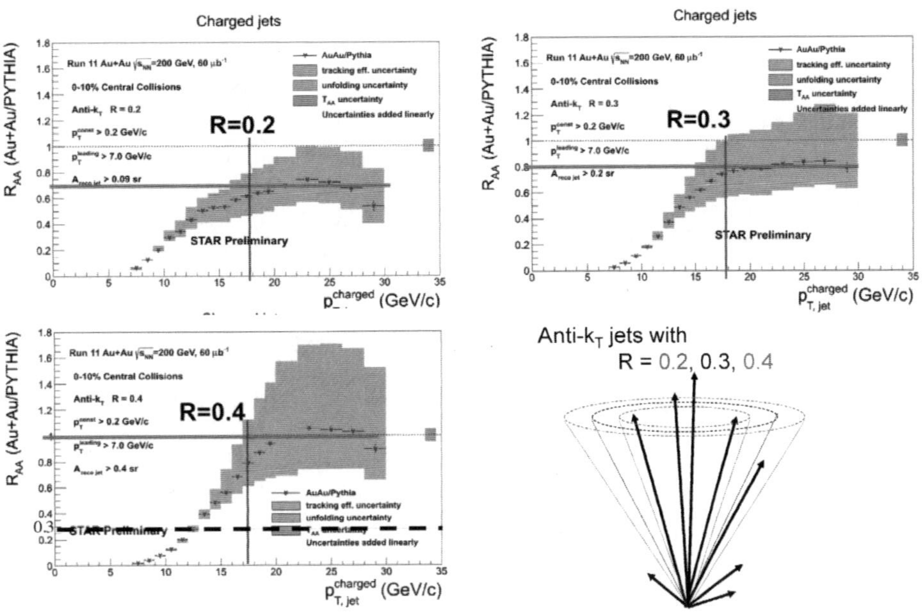

Figure 6. STAR R_{AA} for charged jets at $\sqrt{s_{NN}}$ =200 GeV in central Au+Au collisions for 3 different jet cones with $R = 0.2, 0.3, 0.4$ (see details in legend and sketch) [1].

The good news for the future is that a new detector, now called sPHENIX, to find jets by the more traditional method using hadron calorimetry has been proposed, is moving along on the approval process and is on the schedule at RHIC for partial commisioning in 2019.

4. Kari Eskola once asked me whether I believed in QCD

In the 4th meeting in this series, in Prague in 2009, Kari Eskola asked me whether I believed in QCD after I expressed doubt about some calculation. I answered, "Of course I believe in QCD; but I am skeptical of many calculations that claim to be QCD." Such calculations are still being made which I learned about by reading Jan Rak's talk at a recent conference [30].

4.1. Another wrong calculation claiming to be QCD

Figure 7a [31] shows a supposed QCD calculation of the inclusive jet cross section in $p + p$ collisions at $\sqrt{s_{NN}}$ =7–33 TeV in which the integrated inclusive jet cross section exceeds the inelastic cross section. This is normal for inclusive measurements, e.g. single particle spectra, where the integral of the inclusive cross section equals the interaction cross section times the mean multiplicity, but is well known not to happen in hard-scattering. Nature (i.e. non-perturbative QCD) finds a way to stop the p_T^{-n} divergence, which flattens for $p_T \lesssim 3$ GeV/c as shown for direct-γ production in Fig 7b. The same flattening happens for the p_T distribution of Drell-Yan lepton pair production [32]. Even though the authors of Pythia provided "a phenomenological modification of the low-p_T behiavior of the jet cross section" to agree with

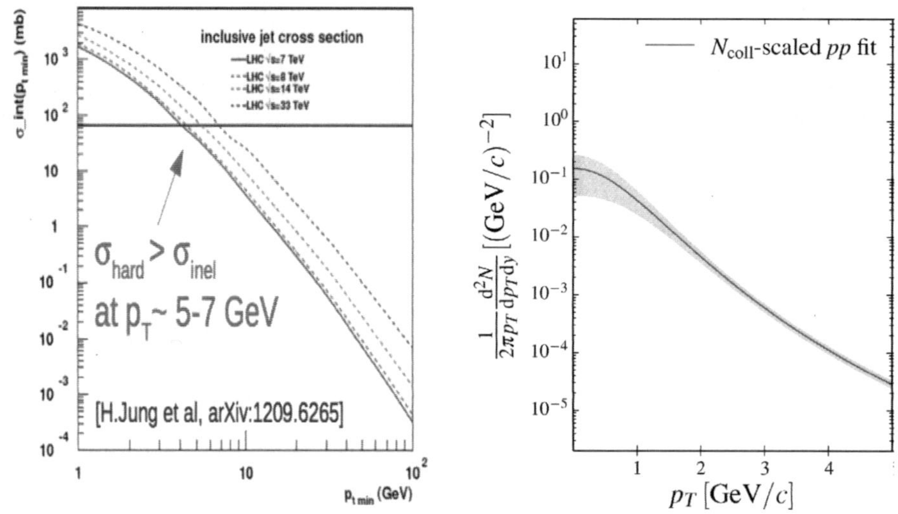

Figure 7. a) (left) Integrated inclusive jet cross section for $p_T \geq p_{T_{\rm min}}$ as a function of $p_{T_{\rm min}}$ from Ref. [31] for several $\sqrt{s_{NN}}$ indicated. b) (right) $(Ed^3\sigma/dp^3)/(0.054$ mb) for direct-γ production in $p+p$ collisions at $\sqrt{s}$ =200 GeV. The distribution is scaled for comparison to Au+Au central (0-20%) measurements [33].

the actual QCD behavior, the Pythia 'calculators' [31] decided not to use it and got a ridiculous answer, once again confirming my response to Kari.

4.2. Direct-γ production, real QCD calculations and x_T scaling

My favorite QCD reaction is direct-γ production via the subprocess $g + q \rightarrow \gamma + q$. This is much better than jet production to test QCD calculations as well as to measure parton energy loss in the QGP for several reasons: i) the γ participates directly in the hard-scattering and then emerges freely and unbiased from the reaction, with no accompanying particles, and passes unaffected through the medium to a detector where its energy can be measured precisely; ii) the transverse momentum of the jet from the outgoing quark at the reaction point is equal and opposite to that of the γ, thus is also precisely known (modulo k_T); iii) for LO pQCD calculations of the direct-γ inclusive spectrum, no fragmentaton functions are needed—a major advantage over jet and single particle calculations. This is illustrated in Fig. 8 where x_T scaling is presented for both inclusive direct-γ (Fig. 8a) over a large range of $\sqrt{s}$ in $p+p$ and $\bar{p}+p$ collisions and for inclusive charged particles at 3 values of $\sqrt{s}$ (Fig. 8b).

x_T scaling [36, 37] provides a totally data driven test of whether pQCD or some other underlying subprocess is at work, without the need to know the details of the structure functions, fragmentation function and coupling constant, as well as providing a compact quantitative way to describe the data using the effective index, $n_{\rm eff}(x_T, \sqrt{s})$. The invariant cross section for inclusive single particle production can be written as:

$$E\frac{d^3\sigma}{dp^3} = \frac{d^3\sigma}{p_T dp_T dy d\phi} = \frac{1}{p_T^{n_{\rm eff}(x_T,\sqrt{s})}} F(x_T) = \frac{1}{\sqrt{s}^{\,n_{\rm eff}(x_T,\sqrt{s})}}\, g(x_T) \quad , \qquad (6)$$

where $Ed^3\sigma/dp^3 = \sigma^{\rm inv}(p_T, \sqrt{s})$ is the invariant cross section for inclusive particle production with transverse momentum p_T at c.m. energy $\sqrt{s}$, and $x_T = 2p_T/\sqrt{s}$. It is important to

IOP Conf. Series: Journal of Physics: Conf. Series **805** (2017) 012005 doi:10.1088/1742-6596/805/1/012005

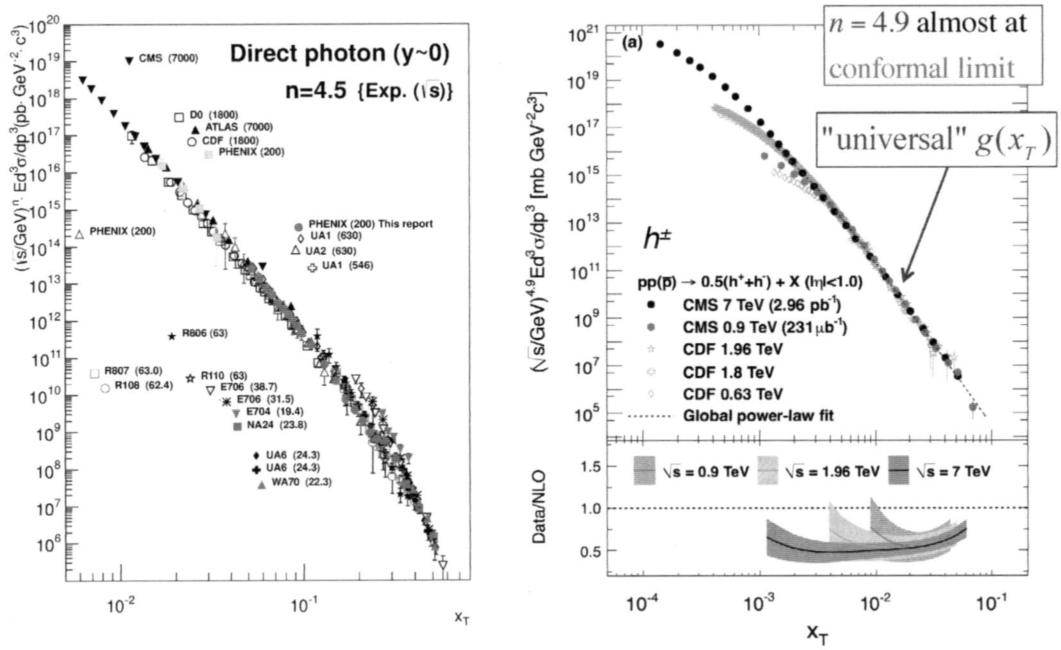

Figure 8. a) (left) Direct-γ measurements plotted as $\sqrt{s}^{n_{\rm eff}} \times Ed^3\sigma/dp^3$ at $x_T \equiv 2p_T/\sqrt{s}$ with $n_{\rm eff} = 4.5$ [34]. The legend gives the experiment and $\sqrt{s}$. b)(right) x_T scaling for inclusive charged particles at $\sqrt{s} \gtrsim 1$ TeV with $n_{\rm eff} = 4.9$ [35].

emphasize that the effective power, $n_{\rm eff}(x_T, \sqrt{s})$, is different from the power n of the invariant cross section, which varies with $\sqrt{s}$ (which it must if x_T scaling is to hold). For pure vector gluon exchange, or without the evolution of α_s and the structure and fragmentation functions in QCD, $n_{\rm eff} = 4$ as in Rutherford scattering. However, due to the non-scaling in QCD [37], the measured value of $n_{\rm eff}$ depends on the x_T value and the range of $\sqrt{s}$ used.

The point of this discussion and Fig. 8 is that the direct-γ data are very well described by QCD and x_T scaling, with $n_{\rm eff} = 4.5$ due to the QCD evolution, while the charged particle data also follow x_T scaling very well, but with a larger $n_{\rm eff} = 4.9$ due to the added non-scaling of the fragmentation functions. This shows that the charged particle cros-sections follow QCD even though the NLO QCD calculations miss the data by a factor of 2, which [35] "suggests that the fragmentation functions are not well tuned for LHC energies."

5. Problems with centrality for reactions with very large p_T in p+A collisions

Last year [29], I discussed a problem (or excitement for some people) with determining the centrality in d+Au at RHIC, using Beam-Beam counters at forward rapidity, $3.1 < \eta < 3.9$, for reactions with very large $p_T > 10$ GeV/c ($x_T > 0.1$) at mid-rapidity. This year, similar methods at LHC for p+Pb collisions at $\sqrt{s_{NN}} = 5.02$ TeV, produce a similar problem at the same x_T (Fig. 9a) [38]. Figure 9b [38] shows that at both LHC and RHIC, avoiding centrality cuts by using minimum bias collisions to measure $R_{p{\rm A}}(p_T) = {\rm A}^{\alpha(p_T)-1}$ gives more reasonable results.[4] This is the basis for the p+A run at RHIC in 2015, using a few values of A to determine $\alpha(p_T)$ of minimum bias p+A collisions rather than make centrality cuts.

[4] I use the original Cronin terminology [5] here where the cross section for hard scattering in p+A collisions was represented as $\sigma_{p{\rm A}}(p_T) = {\rm A}^{\alpha(p_T)} \times \sigma_{pp}(p_T)$, where $\alpha(p_T) \equiv 1$ for pure point-like scattering with no shadowing.

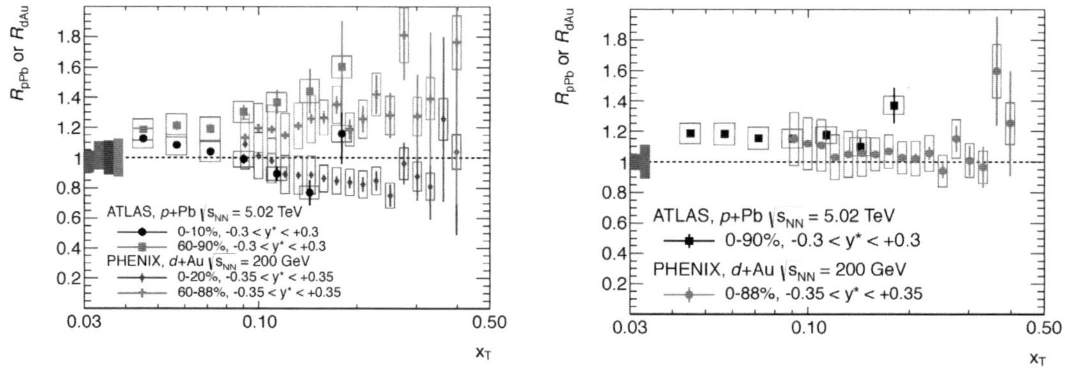

Figure 9. a) (left) $R_{\rm pPb}$ (LHC) and $R_{\rm dAu}$ (RHIC) for jets [38], at $\sqrt{s_{NN}}$ and centralities indicated. b)(right) Same for minimum bias collisions.

6. The importance of $p+p$ comparison data at the same $\sqrt{s}$ in the same detector

Two LHC experiments presented measurements this year of $R_{\rm pPb}$ from p+Pb collisions at $\sqrt{s_{NN}}$ =5.02 TeV from the run in 2013. It was an impressive tour-de-force for the LHC to collide particles with different Z/A in a single ring; but the price the experimenters paid was that they had no comparison $p+p$ data at the same $\sqrt{s}$. The results from the ALICE and CMS experiments are shown in Fig. 10. The ALICE results show $R_{\rm pPb} = 1$, constant for $10 \leq p_T \leq 50$

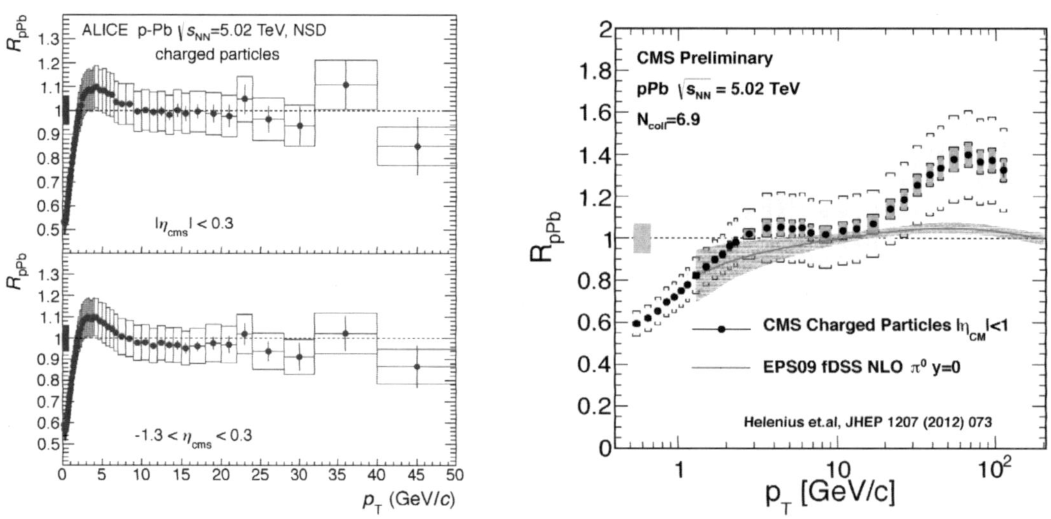

Figure 10. a) (left) ALICE $R_{\rm pPb}$ vs p_T [39] b)(right) CMS preliminary $R_{\rm pPb}$ [40].

GeV/c, while the CMS results agree for $3 \leq p_T \leq 20$ GeV/c, with $R_{\rm pPb} = 1$, but then show a sharp increase to $R_{\rm pPb} \approx 1.4$ for $40 \leq p_T \leq 100$ GeV/c, a jump never before seen in such measurements. For comparison the ATLAS jet measurement at $x_T \geq 0.045$ ($p_T \geq 113$ GeV/c) (Fig. 9b) is constant at $R_{\rm pPb} \approx 1.2 \pm 0.1$. Since there is no $p+p$ comparison measurement for single inclusive particles at $\sqrt{s}$ =5.02 TeV, experience suggests that this is the problem, which must be resolved by a high priority $\sqrt{s}$ =5.02 TeV $p+p$ comparison run when the LHC starts up again.

IOP Conf. Series: Journal of Physics: Conf. Series **805** (2017) 012005 doi:10.1088/1742-6596/805/1/012005

There were similar "exciting results" at CERN in 1982 which had unexpected consequences.

6.1. Experience is the best teacher. Right?
In 1984, a program of Heavy Ions in the CERN-SPS was approved by the DG, Herwig Schopper, partly due to some "exciting results" from $\alpha+\alpha$ collisions in the CERN-ISR (Fig 11a) [41]. The large value of the $\alpha\alpha/pp$ cross sections in Fig. 11a was WRONG because of an incorrect

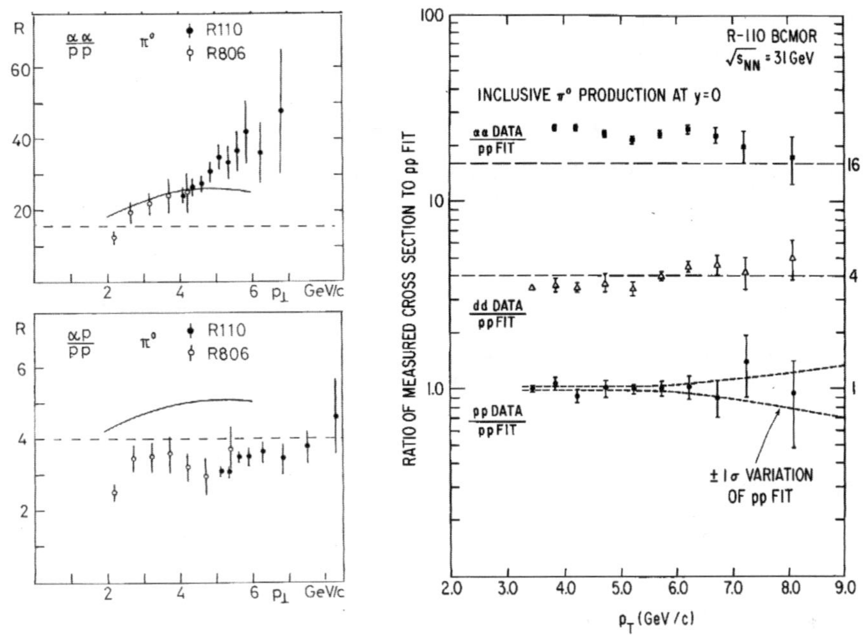

Figure 11. a) (left) Ratio of cross sections in $\alpha+p$ and $\alpha+\alpha$ interactions to the cross sections in $p+p$ interactions as a function of p_T: $R[(\alpha p \to \pi^0+X)/(pp \to \pi^0+X)]$ at $\sqrt{s_{NN}} = 44$ GeV and $R[(\alpha\alpha \to \pi^0+X)/(pp \to \pi^0+X)]$ at $\sqrt{s_{NN}} = 31$ GeV [42], compiled by Faessler [41]. b) (right) BCMOR measurements [43] of the inclusive π^0 cross sections in $\alpha+\alpha$, $d+d$ and $p+p$ collisions at $\sqrt{s_{NN}} = 31$ GeV divided by a fit to the $p+p$ data.

extrapolation of $p+p$ measurements from $\sqrt{s}$ =62.4 to 31 GeV. I complained about this but I was too busy making magnets at ISABELLE at that time—a lucky break in retrospect. Also, because ISABELLE was cancelled in 1983 and the chair of my department, Arthur Schwartzschild, was a nuclear physicist who had heard of this "exciting result" by the grapevine and wanted to get collider experience for the RHIC proposal, he offered me a small group of nuclear physicists to participate in the 1983 CERN-ISR $p+p$, $d+d$ and $\alpha+\alpha$ run at $\sqrt{s_{NN}}$ =31 GeV (the BCMOR collaboration where B stands for Brookhaven). The correct results are shown in Fig. 11b [43].

This shows that sometimes WRONG RESULTS can have a bigger impact than correct results because they are EXCITING; but this does not excuse making mistakes.

Acknowledgments
I would like to acknowledge the fantastic effort by the staff of SUBATECH, the conference organizers, especially Magali Estienne, and the Nantes Police, in finding and returning my wallet totally intact with cash, credit cards and passport, which I had lost leaving the Workshop.

IOP Conf. Series: Journal of Physics: Conf. Series **805** (2017) 012005 doi:10.1088/1742-6596/805/1/012005

References

[1] Lee Y-J 2014 Experimental results on jets in ultra-relativistic nuclear collisions. Presented at *Quark Matter 2014* Darmstadt, Germany, May, 23 2014.
[2] Shuryak E 2013 Comment to talk of M. J. Tannenbaum [29].
[3] Sakaguchi T 2014 Detailed study of parton energy loss via measurement of fractional momentum loss of high p_T hadrons in heavy ion collisions. These proceedings.
[4] Baier R, Schiff D and Zakharov B G 2000 *Annu. Rev. Nucl. Part. Sci.* **50** 37–69
[5] Cronin J W, Frisch H J, Shochet M J, Boymond J P, Piroué P A and Sumner R L 1975 *Phys. Rev.* D **11** 3105–3123
[6] Baier R, Dokshitzer Y L, Mueller A H, Peigné S and Schiff D 1997 *Nucl. Phys.* B **483** 291–320
[7] Gyulassy M and Plümer M 1990 *Phys. Lett.* B **243** 432–438
[8] Wang X-N and Gyulassy M 1992 *Phys. Rev. Lett.* **68** 1480–1483
[9] Thoma M H and Gyulassy M 1991 *Nucl. Phys.* B **351** 491–506
[10] Gyulassy M, Plümer M, Thoma M and Wang X-N 1992 *Nucl. Phys.* A **538** 37c–50c
[11] Gyulassy M and Wang X-N 1994 *Nucl. Phys.* B **420** 583–614
[12] Bernard V, Capella A, Geist W, Gorodetsky P, Seltz R and Voltolini C (eds) 1991 *Quark Gluon Plasma Signatures* (91192 Gif-sur-Yvette Cedex - France: Editions Frontières) URL `http://inspirehep.net/record/967473?ln=en`
[13] ALICE Collaboration 1995 Technical Proposal for A Large Ion Collider Experiment at the CERN-LHC CERN-LHCC-95-71 URL `https://cds.cern.ch/record/293391/files/cer-000214817.pdf`
[14] Matsui T and Satz H 1986 *Phys. Lett.* B **178** 416–422
[15] Baier, R also Tannenbaum, M 1998 *Proc. IV Workshop on Quantum Chromodynamics, Paris, France, 1–6 June* ed Fried H M and Müller B (World Scientific)
[16] Adler S S and others (PHENIX Collaboration) 2005 *Phys. Rev.* C **71** 034908
[17] Rak J and Tannenbaum M J 2013 *High p_T Physics in the Heavy Ion Era* (Cambridge University Press)
[18] Tannenbaum M J 2014 *Int. J. Mod. Phys.* A **29** 1430017 *Preprint* arXiv:1406.1100
[19] Burke K M and others (JET Collaboration) 2014 *Phys. Rev.* C **90** 014909
[20] Baier R, Dokshitzer Y L, Mueller A H, Peigné S and Schiff D 1997 *Nucl. Phys.* B **484** 265–282
[21] Adler S S and others (PHENIX Collaboration) 2006 *Phys. Rev.* D **74** 072002
[22] Adler S S and others (PHENIX Collaboration) 2006 *Phys. Rev.* C **73** 054903
[23] Adcox K and others (PHENIX Collaboration) 2002 *Phys. Rev. Lett.* **88** 022301
[24] Adler S S and others (PHENIX Collaboration) 2007 *Phys. Rev.* C **76** 034904
[25] Adare A and others (PHENIX Collaboration) 2013 *Phys. Rev.* C **88** 024906
[26] Aamodt K and others (ALICE Collaboration) 2011 *Phys. Rev. Lett.* **106** 032301
[27] Alver B and Roland G 2010 *Phys. Rev.* C **81** 054905
[28] Adamcyzk L and others (STAR Collaboration) 2014 *Phys. Rev. Lett.* **112** 122301
[29] Tannenbaum M J 2013 How do quarks and gluons lose energy in the QGP? *Preprint* arXiv:1404.6232 to appear in *Proc. 9th Intl. Workshop on High-pT Physics at LHC, LPSC Grenoble, France, 24-28 September 2013.*
[30] Rak J 2014 High-p_T probes of excited nuclear medium in the LHC era. Presented at *3rd. Int'l. Conf. on New Frontiers in Physics* Kolymbari, Crete, Greece, August 4, 2014.
[31] Grebenyuk A, Hautmann F, Jung H, Katsas P and Knutsson A 2012 *Phys. Rev.* D **86** 117501
[32] Ito A S and others (CFS Collaboration) 1981 *Phys. Rev.* D **23** 604–633
[33] Adare A and others (PHENIX Collaboration) 2014 Centrality dependence of low-momentum direct-photon production in Au+Au collisions at $\sqrt{s_{NN}} = 200$ GeV. Submitted to *Phys. Rev.* C, arXiv:1405.3940.
[34] Adare A and others (PHENIX Collaboration) 2012 *Phys. Rev.* **D 86** 072008
[35] Chatrchyan S and others (CMS Collaboration) 2011 *JHEP* **08** 086
[36] Blankenbecler R, Brodsky S J and Gunion J F 1972 *Phys. Lett.* B **42** 461–465
[37] Cahalan R F, Geer K A, Kogut J and Susskind L 1975 *Phys. Rev.* D **11** 1199–1212
[38] Perepelitsa D V 2013 Inclusive jet production in ultrarelativistic proton-nucleus collisions. Ph. D. Thesis, Columbia University, November, 2013.
[39] Abelev B and others (ALICE Collaboration) 2014 *Eur. Phys. J.* C **74** 3054
[40] Appelt E and others (CMS Collaboration) 2014 *Nucl. Phys.* A **931** 377–381
[41] Faessler M A 1981 *Proc. 1st Intl. Conf. on Physics in Collision (Blacksburg, VA, May 28–31, 1981)* ed Bellini G and Trower W P (New York: Plenum)
[42] Angelis A L S and others (COR Collaboration) 1982 *Phys. Lett.* B **116** 379–382
[43] Angelis A L S, Chasman C, Haustein P E, Olness J W, Tanaka M and others (BCMOR Collaboration) 1987 *Phys. Lett.* B **185** 213–217

IOP Conf. Series: Journal of Physics: Conf. Series **805** (2017) 012006 doi:10.1088/1742-6596/805/1/012006

Multiple Scattering with fully coherent scattering in pA and AA collisions

Haitham Zaraket

Laboratoire de Physique et d'électronique, Lebanese University, Faculty of Sciences (I), Hadath, Beirut Lebanon

E-mail: hzaraket@ul.edu.lb

Abstract. A new energy loss mechanism model is presented. Medium multiple scattering is modified by the interaction with a classical non-abelian background field. It is shown that the obtained gluon radiation spectrum has a symmetric structure in x (the outgoing radiation energy over incident parton energy) and $(1-x)$ and the appropriate sign.

1. Introduction

The medium produced in heavy ion (AA) collision or in pA collision is hard to probe directly. A fast parton produced inside or outside (but passing through) the medium can witness and probe the secrets of the medium and carry it to the outside world (our detectors) in a jet form. Energy loss is among the main processes that the parton will suffer while passing through the medium. This will modify (quench) the the jet.

Energy loss of a fast parton is currently considered as one of the key characteristics of the medium. To decode the message sent by the medium we have to model all the loss mechanisms and determine the dependence of energy loss on the size of the medium (length L) and on the energy (E) of the propagating /parton.

2. Energy loss mechanisms

A simplified model of the propagating medium is a finite sized medium with scattering centers. The main energy loss mechanisms are then collisional and radiation energy losses. However, if one follows the history of the formed medium, it is possible to have a remnant background field. It is believed that during and after relativistic heavy ion collisions, strong chromomagnetic field will form and it can be treated as a classical background. Numerical solutions [1, 2] indicate that "just" after collision transverse color electric and color magnetic fields change suddenly from being transverse in the initial state, in the so-called Color-Glass-Condensate state, to longitudinal. The latter is called glasma flux tubes. Transverse fields then rise, and at some stage after the collision the transverse and longitudinal components of color electric and color magnetic fields reach "steady" comparable values. For gluon radiation with long formation length it is possible that trajectory of the particle is deviated by not only by multiple scattering but also by the the strong magnetic field. In a recent work of Peigné *et al* [3], the authors considered a scenario where the incoming high energy (E) parton undergoing a hard, small angle, scattering in the medium (a non-abelian possibility) with soft re-scattering. It was found

IOP Conf. Series: Journal of Physics: Conf. Series **805** (2017) 012006 doi:10.1088/1742-6596/805/1/012006

that the medium-induced radiative energy loss is proportional to E. The main contribution to medium-induced radiation spectrum arises from the interference between initial and final state radiation. In our model the small angle (hard scattering) is replaced by the effect of the passage through a region where a strong background field is present. The method can be applied easily to:

- Zero background field,
- an impulse field, a field that is nonzero in a very small region to mimic the single hard scattering in [3],
- and a constant field.

In the paper we will focus on the impulse field case.

3. Hard scattering, the linear energy dependence

Considering the propagation of a fast parton in a medium of momentum p, the parton can radiate a gluon of momentum k with energy fraction x. If the formation time of the radiated gluon is large compared to the size of the medium, a fully coherent scattering event will occur where hard small angle scattering is possible besides multiple soft scatterings as shown in Figure 1. In the region where $x \equiv k^+/p^+ \ll 1$ one recovers the Gunion-Bertsch spectrum [4]. The final

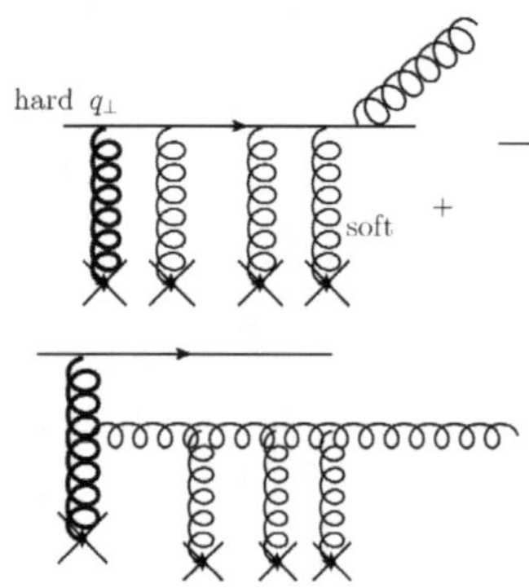

Figure 1. Hard scattering and multiple soft scatterings, three possible diagrams.

spectrum was found [3] to have the linear energy dependence

$$\Delta E \sim (\cdots)E\,.$$

This demonstrates that the energy loss in a finite sized medium is not bounded at high energy. The induced spectrum arises from the interference between (in a time/path ordered approximation) diagrams where the gluon is emitted by the initial parton (before the medium) and that where gluon is radiated from the final state parton (after the medium).

4. Energy loss: synchrotron plus collision

The new proposed mechanism is a modification of the parton "trajectory" by the background field while suffering multiple soft scatterings. By "tuning" the strength and the space-time dependence of the field it is possible to mimic the hard scattering presented in the last section.

Using the light cone formalism proposed by Zakharov [5] it is possible to write the gluon emission probability as

$$\frac{dP}{dx} = \int_0^L dz n(z) \frac{d\sigma_{eff}(x,z)}{dx} \tag{1}$$

where $n(z)$ is the density of the scattering centers in the medium. The effective cross section, with transverse variable $\boldsymbol{\rho}$, is approximated in the light cone approximation by

$$\frac{d\sigma_{eff}}{dx} = \mathrm{Re} \int d\boldsymbol{\rho} \psi^*_m(\boldsymbol{\rho}, x)\sigma_3(\boldsymbol{\rho}, x)\psi_f(\boldsymbol{\rho}, x, z) \ . \tag{2}$$

σ_3 is the cross section of interaction of the $qg\bar{q}$ system, the analogue of the three diagrams in the Gunion-Bertsch approximation (Figure 1), ψ_f is the light-cone wave function in the background field, ψ_m is the light-cone medium modified wave function, given by

$$\psi_f(\boldsymbol{\rho}, x) = P(x)\left(\frac{\partial}{\partial \rho'_x} - i\lambda_g \frac{\partial}{\partial \rho'_y}\right)\int_0^\infty d\xi \mathcal{K}_f(\boldsymbol{\rho}, z|\boldsymbol{\rho}', z-\xi)\Bigg|_{\rho'=0} \tag{3}$$

$$\psi^*_m(\boldsymbol{\rho}, x, z) = -P(x)\left(\frac{\partial}{\partial \rho'_x} - i\lambda_g \frac{\partial}{\partial \rho'_y}\right)\int_0^\infty d\xi \mathcal{K}_m(\boldsymbol{\rho}, z+\xi|\boldsymbol{\rho}', z)\Bigg|_{\rho'=0} \tag{4}$$

where $P(x) = i\sqrt{\alpha_s/2x}[2\lambda_q x + \lambda_g(2-x)]/2\mu(x)$. The light cone propagators K_f and K_m are obtained by solving the problem of a fast parton in a background field and in medium respectively. Hence the emission probability is the interference between medium modified and background modified wave functions.

To mimic the hard scattering we consider a background field of the form

$$H = H_o\delta(z - z_s)$$

which allows us to model the hard scattering event. The propagator K_f is evaluated exactly while the propagator K_m is approximated by the $N = 1$ term. The approximated spectrum is then written as

$$\begin{aligned} x\frac{dP}{dx} &= C_F[1 + (1-x)^2 - \frac{x^2}{N_c^2}][1 - x + \frac{x^2}{2}]\frac{\alpha_s}{\pi} \\ &\times \frac{L}{\lambda_g}\log\left(1 + \frac{\mu_s^2}{(\mathbf{q}'_\perp)^2(q'x - (1-x)Q)^2}\right) \end{aligned} \tag{5}$$

where q' and Q are the color charge of the outgoing quark and gluon respectively and the gluon mean free path λ_g is defined as

$$\frac{1}{\lambda_g} = \frac{\alpha_s^2 \pi C_A C_T n}{2\mu_s^2} \ .$$

The present form, which is symmetric in "x", $1 - x$, is expected in a process where the incident parton is a gluon. For a propagating gluon, the radiated gluon and the outgoing gluon are expected to be symmetric in energy repartition.

Eq. (5) should be compared with

$$x\frac{dP}{dx} = (2C_R - N_c)\frac{\alpha_s}{\pi}\frac{L}{\lambda_g}\log\left(1 + \frac{\mu_s^2}{(\mathbf{q}_\perp)^2 x^2}\right) \tag{6}$$

obtained in [3] for small x. The main concern with eq. (6) is that the color factor $(2C_R - N_c)$ is $-1/N_c$ for a propagating quark, this negative result was interpreted as a phase space artifact. Our result is perfectly positive and hence no potential energy gain mechanism was found. To understand the contracting results it is important to outline the main differences between the two proposed energy "loss" mechanisms:

- In this article induced gluon radiation was defined as the radiation spectrum of a parton in a medium plus background minus the radiation of the parton in background field. While in [3] the induced gluon radiation spectrum was defined as that of a parton in a medium minus the vacuum contribution.
- A full x treatment was done in the present article, while small x approximation is done in [3]. However if one considers the small x limit of eq. (5) the result is still positive.

Finally we integrate over x , to get the total energy loss in the two models to obtain

$$\Delta E \equiv E \int dx x \frac{dP}{dx} \approx \mathcal{C}\alpha_s \frac{L}{\lambda_g} \frac{2\mu}{q q'_\perp} E \, . \tag{7}$$

In the background approach the color factor $\mathcal{C}$ contains detailed information about the color charge of the incident quark, final quark and the radiated gluon, while in the hard scattering approach [3] $\mathcal{C}$ is $2C_R - N_c$. Hence in both case energy loss is linear in E.

5. Conclusion

We have proposed a model for the propagation of an asymptotic parton in a medium where besides scattering a remanent non-abelian background field is also present. In the impulse field approximation the effect of the background can be related to the proposed model by Peigné *et al* where multiple soft scatterings are coupled to hard scattering. The two results were found to have opposite signs, which both agree on the final E dependence.

Acknowledgments

The work of H.Z. was supported by the French CNRS through the PICS project "Parton production and propagation in hot QCD matter" and partially by the Lebanese university. He would also like to thank his colleagues at the LPSC for their hospitality. He also thanks Stephane Peigné for useful discussion.

References

[1] Fries R J, Kapusta J I and Li Yang 2006 From color fields to Quark Gluon Plasma, *Nucl. Phys.* **A 774**, 861
[2] Lappi T, and McLerran L 2006 Some features of the glasma *Nucl. Phys.* **A 772**, 200
[3] Peigné S, Arleo F and Kolevatova R 2014 Medium-induced gluon radiation: an update *Preprint* hep-ph/1402.1671v1
[4] J.F. Gunion and G. Bertsch 1982 Phys. Rev. **D25**, 746
[5] Zakharov B.G. 1997 Radiative energy loss of high-energy quarks in finite-size nuclear matter and quark-gluon plasma, *JETP Lett.* **65**, 615.

IOP Conf. Series: Journal of Physics: Conf. Series **805** (2017) 012007 doi:10.1088/1742-6596/805/1/012007

Jet fragmentation and multijet studies in heavy ion collisions at ATLAS

Martin Spousta (*for the ATLAS Collaboration*)
Charles University in Prague, Institute of Particle and Nuclear Physics,
V Holesovickach 2, 180 00 Prague 8, Czech republic
E-mail: martin.spousta@cern.ch

Abstract. We report on two measurements done by ATLAS in a context of exploring the jet quenching physics at LHC: the jet fragmentation measurement and the measurement of neighbouring jet production. Both of these measurements were done using the data collected in 2011 Pb+Pb run providing Pb+Pb collisions at $\sqrt{s_{\mathrm{NN}}} = 2.76$ TeV. These measurements are expected to provide an insight into the modifications of jets initiated by partons passing through hot and dense QCD matter created in heavy ion collisions.

1. Introduction

In this short report, we summarize two recent measurements done by ATLAS related to the jet quenching: the measurement of jet fragmentation [1] and the measurement of neighbouring jet production [2]. Both of these measurements are expected to improve our understanding of jet quenching. Jet quenching refers, collectively, to a set of possible modifications of parton showers by the quark-gluon plasma through interactions of the constituents of the shower with the colour charges in the plasma [3]. Shower constituents my be elastically or inelastically scattered resulting in both deflection and energy loss of the constituents of the shower. This modification of jet internal structure may be then detected as a modification of the jet fragmentation functions. If the parton shower is modified in its early stage when hard gluons are radiated, then rates of neighbouring jets may also be affected. The neighbouring jets originating from the same hard interaction propagate in approximately the same direction and should have similar path lengths in the medium. Therefore, measuring the neighbouring jet production may probe the differences in the quenching of two jets that do not result primarily from the difference in the path length.

2. Data selection

The measurements summarized in this report use the data collected in 2011 Pb+Pb run providing Pb+Pb collisions at $\sqrt{s_{\mathrm{NN}}} = 2.76$ TeV. The data were selected using high-level trigger (HLT) seeded by a Level-1 minimum-bias trigger. The Level-1 trigger required a total transverse energy measured in the calorimeter greater than 10 GeV. The HLT jet trigger required a presence of underlying-event-subtracted jet with transverse energy greater than 20 GeV. The total integrated luminosity of data sample selected by the HLT trigger is 0.14 nb^{-1}.

The measurement of neighbouring jets uses also the data selected by the minimum bias (MB) trigger. MB triggers use the zero degree calorimeter (ZDC) covering the pseudorapidity range of $|\eta| > 8.3$, the inner detector (ID) covering $|\eta| < 2.1$, and minimum-bias trigger scintilators

IOP Conf. Series: Journal of Physics: Conf. Series **805** (2017) 012007 doi:10.1088/1742-6596/805/1/012007

(MBTS) covering the pseudorapidity range of $2.1 < |\eta| < 3.9$. The total integrated luminosity of data sample triggered by MB trigger is $7\mu b^{-1}$.

The reference Monte Carlo (MC) event sample was obtained by overlaying simulated PYTHIA [4] pp hard-scattering events at $\sqrt{s} = 2.76$ TeV onto 1.2 million minimum-bias Pb+Pb events recorded in 2011.

The centrality of Pb+Pb collisions was characterized by $\Sigma E_{\mathrm{T}}^{\mathrm{FCal}}$, the total transverse energy measured in the forward calorimeters. The measurement of fragmentation functions was performed in seven centrality bins defined according to successive percentiles of the $\Sigma E_{\mathrm{T}}^{\mathrm{FCal}}$ distribution ordered from the most central to the most peripheral collisions: 0–10%, 10–20%, 20–30%, 30–40%, 40–50%, 50–60%, and 60–80%. The measurement of neighbouring jet production was performed in four centrality bins, 0–10%, 10–20%, 20–40%, and 40–80%. The most peripheral bin was used in each measurement to evaluate the ratios of measured distributions in order to quantify the suppression of yields in central with respect to peripheral collisions.

Jets were reconstructed using the anti-k_t jet finding algorithm [5] with three different distance parameters, $R = 0.2$, $R = 0.3$, and $R = 0.4$. The detailed description of the jet reconstruction and the algorithm for the UE subtraction can be found in Ref.[6].

3. Fragmentation function measurement

The internal structure of the jet was characterized by measuring the transverse momentum ($p_{\mathrm{T}}^{\mathrm{ch}}$) and longitudinal momentum fraction ($z \equiv p_{\mathrm{T}}^{\mathrm{ch}} \cos\Delta R \;/\; p_{\mathrm{T}}^{\mathrm{jet}}$) distributions of charged particles with $p_{\mathrm{T}}^{\mathrm{ch}} > 2$ GeV produced within an angular range $\Delta R = 0.4$ of the reconstructed jet directions. Here, $\Delta R = \sqrt{(\Delta\phi)^2 + (\Delta\eta)^2}$ where $\Delta\phi$ ($\Delta\eta$) is the difference in azimuthal angles (pseudorapidities) between the charged particle and jet directions. Jets were required to have $p_{\mathrm{T}}^{\mathrm{jet}} > 85, 92$, and 100 GeV for $R = 0.2, 0.3$, and 0.4, respectively.

Two different sets of charged-particle fragmentation distributions were measured for each centrality bin and R value, the transverse momentum distribution[1],

$$D(p_{\mathrm{T}}) \equiv \frac{1}{\varepsilon}\left(\frac{1}{N_{\mathrm{jet}}}\frac{\Delta N_{\mathrm{ch}}}{\Delta p_{\mathrm{T}}^{\mathrm{ch}}} - \frac{\mathrm{d}N_{\mathrm{ch}}^{\mathrm{UE}}}{\mathrm{d}p_{\mathrm{T}}^{\mathrm{ch}}}\right), \tag{1}$$

and the fragmentation function,

$$D(z) \equiv \frac{1}{\varepsilon}\left(\frac{1}{N_{\mathrm{jet}}}\frac{\Delta N_{\mathrm{ch}}}{\Delta z} - \left.\frac{\mathrm{d}N_{\mathrm{ch}}^{\mathrm{UE}}}{\mathrm{d}p_{\mathrm{T}}^{\mathrm{ch}}}\right|_{p_{\mathrm{T}}^{\mathrm{ch}} = z p_{\mathrm{T}}^{\mathrm{jet}}}\right). \tag{2}$$

Here N_{jet} represents the total number of jets in a given centrality bin, ΔN_{ch} represents the number of measured charged particles within $\Delta R = 0.4$ of the jets, $1/\varepsilon$ stands for the tracking efficiency correction, and $\mathrm{d}N_{\mathrm{ch}}^{\mathrm{UE}}/\mathrm{d}p_{\mathrm{T}}^{\mathrm{ch}}$ represents the background contribution coming from the underlying event (UE) which is subtracted. The background contribution,

$$\frac{\mathrm{d}N_{\mathrm{ch}}^{\mathrm{UE}}}{\mathrm{d}p_{\mathrm{T}}^{\mathrm{ch}}} = \frac{1}{N_{\mathrm{cone}}}\frac{\Delta N_{\mathrm{ch}}^{\mathrm{cone}}(p_{\mathrm{T}}^{\mathrm{ch}}, p_{\mathrm{T}}^{\mathrm{jet}}, \eta^{\mathrm{jet}})}{\Delta p_{\mathrm{T}}^{\mathrm{ch}}}, \tag{3}$$

is evaluated for each jet radius separately in events having at least one jet above the jet p_{T} thresholds using N_{cone} cones of radius $R = 0.4$ covering the full inner detector by a regular grid.

Any such cone having a charged particle with $p_{\mathrm{T}}^{\mathrm{ch}} > 6$ GeV was assumed to be associated with a real jet in the event and was excluded from the UE background determination. The

[1] Transverse momentum is defined with respect to the beam axis.

IOP Conf. Series: Journal of Physics: Conf. Series **805** (2017) 012007 doi:10.1088/1742-6596/805/1/012007

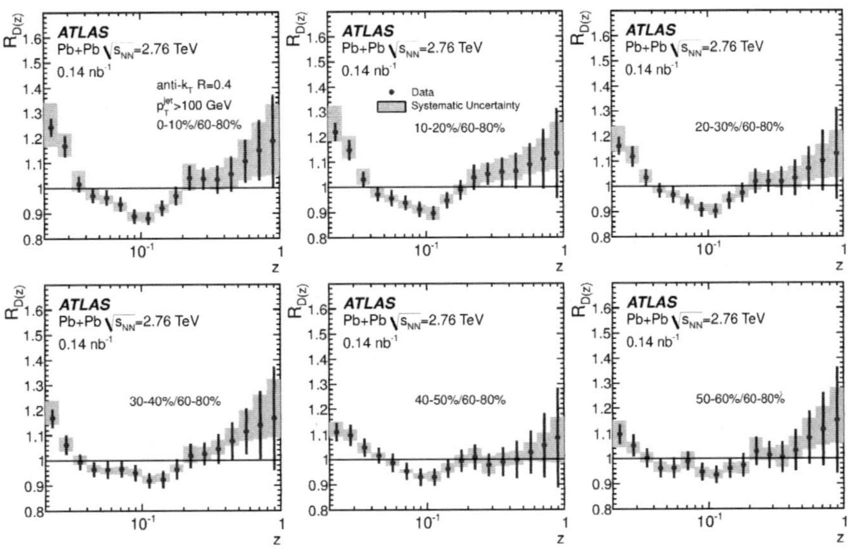

Figure 1. Ratios of unfolded $D(z)$ distributions for six bins in collision centrality to those in peripheral (60–80%) collisions, $D(z)|_{\text{cent}}/D(z)|_{60-80}$, for $R = 0.4$ jets. The error bars on the data points indicate statistical uncertainties while the yellow shaded bands indicate systematic uncertainties.

subtracted background was further corrected to account for proper pseudorapidity distribution of UE and for the elliptic flow.

Two corrections were performed to account for the impact of finite jet energy resolution on the measured distributions. First, the energy of each jet entering the evaluation of $D(p_T)$ and $D(z)$ was shifted to account for the mean effect of the "upfeeding" in a steeply falling jet p_T spectrum produced by the finite jet energy resolution. Then, obtained measured distributions were unfolded to the truth level using SVD – a one-dimensional Singular Value Decomposition (SVD) method [7] implemented in `RooUnfold` [8] which removes also the effects of finite charged particle p_T resolution.

The fragmentation functions are steeply falling distributions. To evaluate the centrality dependence of the fragmentation functions, ratios $R_{D(z)}$ were calculated of the $D(z)$ distributions for all centrality bins excluding the peripheral bin to the $D(z)$ measured in the peripheral, 60–80% centrality bin. The results of $R_{D(z)}$ ratios for $R = 0.4$ jets are shown in Fig. 1. The ratios for all centralities show an enhanced yield of low z fragments ($z \lesssim 0.04$) and a suppressed yield of fragments at intermediate z values ($0.04 \lesssim z \lesssim 0.2$) in more central collisions relative to the 60–80% centrality bin. For the 0–10% centrality bin, the yield of fragments at $z = 0.02$ is enhanced relative to that in the 60–80% centrality bin by 25% while the yield at $z = 0.1$ is suppressed by about 10%. The yield at $z > 0.4$ is enhanced in central with respect to 60–80% peripheral collisions. The size of the observed modifications at low, intermediate, and high z decreases gradually from central to peripheral collisions. The same effects are seen in the ratios of transverse momentum distributions, $R_{D(p_T)}$, as well. The same effects are seen also for $R = 0.3$ and $R = 0.2$ jets. The fluctuations in the UE are approximately 100% (30%) smaller for $R = 0.2$ ($R = 0.3$) jets than they are for $R = 0.4$ jets and, thus, jets reconstructed with smaller jet radii have significantly better jet energy resolution.

To further quantify the effects of the modifications, observed in Fig. 1, on the actual distribution of fragments within the measured jets, the differences in fragmentation functions, $\Delta D(z) = D(z)|_{\mathrm{cent}} - D(z)|_{60-80}$ were calculated and integrals of these distributions, $\int \Delta D(z)\mathrm{d}z$ were evaluated over three z ranges chosen to match the observations: 0.02–0.04, 0.04–0.2, and 0.4–1. The results indicate an increase in the number of particles with $0.02 < z < 0.04$ of less than one particle per jet in the 0–10% centrality bin relative to the 60–80% centrality bin. A decrease of about 1.5 particles per jet is observed for $0.04 < z < 0.2$. The differences between the integrals of the fragmentation functions over $0.4 < z < 1$ are not significant relative to the uncertainties. The results for $\int \Delta D(z)\mathrm{d}z$ further indicate that in the most central collisions, a small fraction, < 2%, of the jet transverse momentum is carried by the excess particles in $0.02 < z < 0.04$ for central collisions, but that the depletion in fragment yield in $0.04 < z < 0.2$ accounts on average for about 14% of $p_{\mathrm{T}}^{\mathrm{jet}}$.

4. Neighbouring jet production

The rate of the neighbouring jets that accompany a test jet, $R_{\Delta R}$, is defined as

$$R_{\Delta R} = \frac{1}{\mathrm{d}N_{\mathrm{jet}}^{\mathrm{test}}/\mathrm{d}E_{\mathrm{T}}^{\mathrm{test}}} \sum_{i=1}^{N_{\mathrm{jet}}^{\mathrm{test}}} \frac{\mathrm{d}N_{\mathrm{jet},i}^{\mathrm{nbr}}}{\mathrm{d}E_{\mathrm{T}}^{\mathrm{test}}}(E_{\mathrm{T}}^{\mathrm{test}}, E_{\mathrm{T,min}}^{\mathrm{nbr}}, \Delta R), \tag{4}$$

where $E_{\mathrm{T}}^{\mathrm{test}}$ and $E_{\mathrm{T}}^{\mathrm{nbr}}$ are the transverse energies of the test and neighbouring jet, respectively; $N_{\mathrm{jet}}^{\mathrm{test}}$ is the number of test jets in a given $E_{\mathrm{T}}^{\mathrm{test}}$ bin and $N_{\mathrm{jet}}^{\mathrm{nbr}}$ is number of neighbouring jets. The measurement of neighbouring jets uses d instead of R to label the distance parameter of jet finding algorithm in order to avoid a confusion with the main measured quantity. For each choice of d, neighbouring jets are considered if they lie within a specific annulus in ΔR away from the test jet: $0.5 < \Delta R < 1.6$, $0.6 < \Delta R < 1.6$, and $0.8 < \Delta R < 1.6$ for $d = 0.2$, $d = 0.3$, and $d = 0.4$ jets, respectively.

The $R_{\Delta R}$ measurement is performed differentially in collision centrality, transverse energy of the test jet, $E_{\mathrm{T}}^{\mathrm{test}}$, and transverse energy of the neighbouring jet, $E_{\mathrm{T}}^{\mathrm{nbr}}$. The raw $R_{\Delta R}$ distributions are corrected for a contribution from neighbouring jets that originate from different hard partonic interaction in the same Pb+Pb collision. The impact of the finite JER, reconstruction efficiency and position resolution on neighbouring jets is corrected for by applying bin-by-bin correction factors in transverse energy to the raw $R_{\Delta R}$ distributions that are derived from the MC data.

Fig. 2 shows the fully corrected $R_{\Delta R}$ distributions for $d = 0.4$ and $d = 0.2$ jets evaluated as a function of $E_{\mathrm{T}}^{\mathrm{test}}$. The distributions are shown for four centrality selections and three selections on minimum transverse energy of neighbouring jets, $E_{\mathrm{T}}^{\mathrm{nbr}} > 30$, 45, and 60 GeV. The $R_{\Delta R}$ distribution exhibits an increase with increasing $E_{\mathrm{T}}^{\mathrm{test}}$, which is consistent in shape with the previous measurement of the same quantity by DØ in $p + \bar{p}$ collisions [9]. Sizable differences between the four different centrality selections are observed for all three jet radii. The yield of neighbouring jets is suppressed as the centrality of the collision increases.

To further quantify the neighbouring jet yields, the per-test-jet normalized E_{T} spectra of neighbouring jets were evaluated,

$$\frac{\mathrm{d}R_{\Delta R}}{\mathrm{d}E_{\mathrm{T}}^{\mathrm{nbr}}} = \frac{1}{\mathrm{d}N_{\mathrm{jet}}^{\mathrm{test}}/\mathrm{d}E_{\mathrm{T}}^{\mathrm{test}}} \sum_{i=1}^{N_{\mathrm{jet}}^{\mathrm{test}}} \frac{\mathrm{d}^2 N_{\mathrm{jet},i}^{\mathrm{nbr}}}{\mathrm{d}E_{\mathrm{T}}^{\mathrm{test}}\mathrm{d}E_{\mathrm{T}}^{\mathrm{nbr}}}(E_{\mathrm{T,min}}^{\mathrm{test}}, E_{\mathrm{T}}^{\mathrm{nbr}}, \Delta R). \tag{5}$$

The trend of the suppression is less pronounced in these steeply falling spectra, thus the differences in the jet spectra evaluated as a function of centrality were quantified by fitting the spectra by a power-law functional form, $\propto 1/{E_{\mathrm{T}}}^n$, and the power index has been extracted

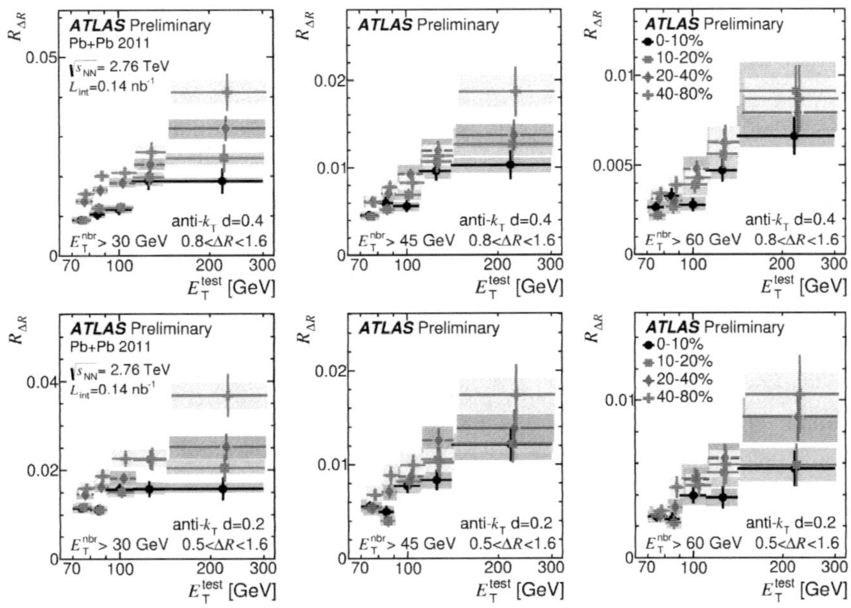

Figure 2. $R_{\Delta R}$ distributions for $d = 0.4$ jets (upper) and $d = 0.2$ jets (lower) evaluated as a function of E_T^{test}. The three different columns show $R_{\Delta R}$ distributions evaluated for three different thresholds on minimum neighbouring jet transverse energy, $E_T^{nbr}> 30$, 45, and 60 GeV. The four different centrality bins are denoted by different markers in each plot. The shaded bands indicate systematic uncertainties, vertical errors represent statistical uncertainties. The data points for the 10–20%, 20–40%, and 40–80% centrality bins are shifted along the horizontal axis with respect to the 0–10% centrality bin for clarity.

for all three choices of jet radius and four centrality bins. The E_T spectra measured in central and peripheral collisions differ in the power-law index by approximately two standard deviations for both $d = 0.4$ and $d = 0.3$ jets, suggesting that the E_T spectra may be less steep in central than peripheral collisions.

To quantify the centrality dependence of the neighbouring jet yields, the ratio $\rho_{R_{\Delta R}}$ is calculated from $R_{\Delta R}$ measured in each centrality bin excluding the peripheral bin to $R_{\Delta R}$ measured in the 40–80%, peripheral, bin. These ratios are evaluated for $d = 0.4$ jets which suffer the least from the statistical uncertainties which are still large. Nevertheless, several characteristic features can be observed. The $\rho_{R_{\Delta R}}$ does not exhibit any strong dependence on E_T^{test}. The suppression factor $\rho_{R_{\Delta R}}$ of the most central collisions is at the level of $0.5 - 0.7$ for all three thresholds on E_T^{nbr}. The suppression becomes less pronounced with decreasing centrality. This is qualitatively consistent with the observation of the centrality dependent suppression of inclusive jet yields [6]. In that measurement, the suppression of the inclusive jet yields was evaluated in terms of the ratio R_{CP} of inclusive jet yields in central to 60–80% peripheral collisions spanning the jet p_T range of $40 - 200$ GeV. Values of $R_{CP} \sim 0.5$ were measured in the 0–10% most central collisions and exhibited only a modest jet p_T dependence. Contrary to a modest dependence of $\rho_{R_{\Delta R}}$ on the test jet E_T, the $\rho_{R_{\Delta R}}$ evaluated as a function of E_T^{nbr} suggests a decrease of suppression with increasing E_T^{nbr}. Such a decrease of suppression with increasing E_T^{nbr} may in fact be expected. The jet quenching is generally expected to depend on the initial parton energy, but if the parton splitting happens such that the two partons have similar energy, their quenching would likely be similar due to similar in-medium path-length traveled by the two partons forming neighbouring jets. Thus, in the configuration of $E_T^{nbr} \approx E_T^{test}$ the per test jet normalization effectively removes the impact of the suppression.

5. Conclusions

In this short report, we have discussed two recent measurements by ATLAS, the jet fragmentation measurement and the measurement of neighbouring jet production. Significant changes in the jet fragmentation have been observed in central with respect to peripheral collisions, namely the enhancement of charged particle yields at low transverse momenta, depletion at intermediate transverse momenta, and an enhancement at high transverse momenta of charged particles.

The production of pairs of correlated jets was quantified using the rate of neighbouring jets that accompany a test jet, $R_{\Delta R}$, evaluated both as a function of test jet E_T and neighbouring jet E_T. A significant dependence of $R_{\Delta R}$ on collision centrality has been observed in both cases, suggesting suppression of neighbouring jets which increases with increasing centrality of the collision reaching a factor of 0.5-0.7 suppression in central with respect to peripheral collisions.

This and other details investigated in these studies implies a complex behaviour of the jet modifications which should provide an important input for theoretical calculations aiming to describe the physics of jet quenching.

Acknowledgments

This work was supported by Charles University in Prague, projects PRVOUK P45, UNCE 204020/2012, and INGO LG13009.

References

[1] ATLAS Collaboration, arXiv:1406.2979.
[2] A. Collaboration, ATLAS-CONF-2014-028. [link].
URL http://cds.cern.ch/record/1703260
[3] G.-Y. Qin, B. Muller, Phys. Rev. Lett. 106 (2011) 162302. arXiv:1012.5280.
[4] T. Sjostrand, S. Mrenna, P. Z. Skands, JHEP 05 (2006) 026. arXiv:hep-ph/0603175.
[5] M. Cacciari, G. P. Salam, G. Soyez, JHEP 04 (2008) 063. arXiv:0802.1189.
[6] ATLAS Collaboration, Phys. Lett. B 719 (2013) 220–241. arXiv:1208.1967.
[7] A. Hocker, V. Kartvelishvili, Nucl. Instrum. Meth. A 372 (1996) 469–481. arXiv:hep-ph/9509307.
[8] http://hepunx.rl.ac.uk/~adye/software/unfold/RooUnfold.html.
[9] V. M. Abazov, et al., Phys. Lett. B718 (2012) 56–63. arXiv:1207.4957.

IOP Conf. Series: Journal of Physics: Conf. Series **805** (2017) 012008 doi:10.1088/1742-6596/805/1/012008

Measurement of inclusive jet spectra in pp, p–Pb, and Pb–Pb collisions with the ALICE detector

Rüdiger Haake for the ALICE Collaboration
Institut für Kernphysik, Westfälische Wilhelms-Universität Münster, 48149 Münster, Germany
E-mail: ruediger.haake@uni-muenster.de

Abstract. Highly energetic jets are sensitive probes of the kinematic properties and the topology of high energy hadron collisions. Jets are collimated sprays of charged and neutral particles, which are produced in fragmentation of hard scattered partons from an early stage of the collision. In ALICE, jets have been measured in pp, p–Pb, and Pb–Pb collisions at several collision energies. While analyses of Pb–Pb events unveil properties of the hot and dense medium formed in heavy-ion collisions, pp and p–Pb collisions can shed light on hadronization and cold nuclear matter effects in jet production. Additionally, pp and p–Pb collisions serve as a baseline for disentangling hot and cold nuclear matter effects. A possible modification of the initial state is tested in p–Pb analyses. For the extraction of a jet signal, the exact evaluation of the background from the underlying event is an especially important ingredient. Due to the different nature of underlying events, each collision system requires a different analysis technique for removing the effect of the background on the jet sample. The focus of this publication is on the ALICE measurements of nuclear modification factors connecting p–Pb and Pb–Pb events to pp collisions. Furthermore, the radial jet structure is explored by comparing jet spectra reconstructed with different resolution parameters.

1. Introduction

Jets can conceptually be described as the final state of scattered partons in a hard scattering. Therefore, jets are an excellent tool to access a very early stage of the collision. The jet constituents represent the final state remnants of the fragmented and hadronized partons that were scattered in the reaction. While all the detected particles have been created in a non-perturbative process (i.e. by hadronization), ideally, jets represent the kinematic properties of the originating partons. Thus, jets are mainly determined by perturbative processes due to the high momentum transfer and the cross sections can be calculated within pQCD. This conceptual definition is descriptive and very simple, the technical analysis of those objects has to be done rigorously though.

In pp collisions, jets can serve as a testing ground of pQCD. In Pb–Pb collisions, the medium effects like those from the quark-gluon plasma (QGP) can be probed. For p–Pb collisions, formation of a hot, dense, and extended medium is not expected and therefore cold nuclear matter effects, e.g. those described by nuclear parton distribution functions (nPDFs), should be dominant.

IOP Conf. Series: Journal of Physics: Conf. Series **805** (2017) 012008 doi:10.1088/1742-6596/805/1/012008

In these proceedings, the focus has been put on the inclusive jet results currently available from the ALICE experiment. Results and conclusions of recent ALICE measurements in pp, p–Pb, and Pb–Pb are presented in Sec. 4. Beforehand, a short description of the ALICE detector with regard to jet measurements is given in Sec. 2. Additionally, the main details of the analyses are shown in Sec. 3.

2. The ALICE detector

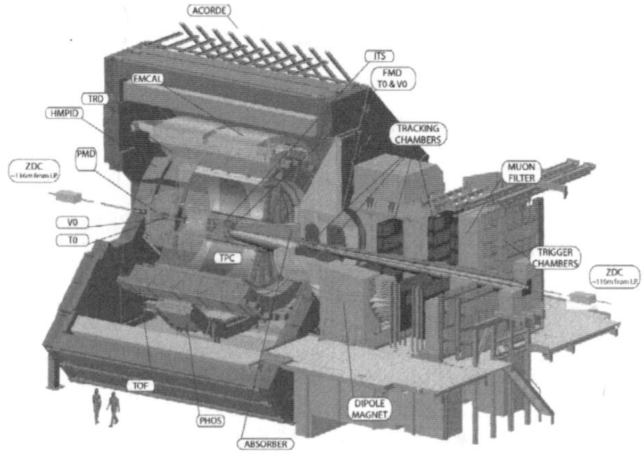

Figure 1. A sketch of the ALICE detector

The ALICE detector (see Fig. 1) is the dedicated heavy-ion experiment at the LHC studying QGP properties and the QCD phase diagram in general. The detector is designed as a general-purpose heavy-ion detector [1] to measure and identify hadrons, leptons, and photons down to very low transverse momenta.

One of ALICE's strengths is the excellent charged particle tracking capability. The tracking is performed in the central barrel, which mainly consists of the Inner Tracking System (ITS) and the Time Projection Chamber (TPC). The ITS [2] is a cylindrical six-layered device consisting of three different semiconductor subdetectors: silicon pixel, drift, and strip detectors (SPD, SDD, and SSD). It directly surrounds the beam pipe. Around the ITS, the Time Projection Chamber is placed. The TPC [3], mainly filled with neon gas at atmospheric pressure, has an external radius of 250cm, the gas detector has a radius of 250 cm and a length of 500 cm. Combining the ITS and the TPC spacepoints, a straight track can be reconstructed within a pseudorapidity interval of $|\eta| < 0.9$.

To measure fully reconstructed jets, also neutral particles have to be detected. For this purpose, a lead-scintillator sampling electromagnetic calorimeter – the EMCal [4] – is used. The calorimeter covers the pseudorapidity interval $|\eta| < 0.7$ and roughly 107° in azimuth. While the EMCal is very sensitive to photons, e.g. from neutral pion decays, the neutron reconstruction capability is less precise.

Finally, the VZERO [5] scintillation counter is used for event characterization and centrality estimation.

3. Jet reconstruction and correction techniques

Jet reconstruction is usually a multistep procedure. At first, beyond the conceptual definition of jets as hadronized partons, a technical jet definition is necessary. There is no unambiguous jet definition and no common way how to measure a jet. The currently most prevalent jet definition at the LHC is given by the anti-k_T algorithm [6] implemented in the FastJet [7] package. For all presented analyses, the anti-k_T algorithm is used to measure signal jets. Additionally, the k_T algorithm is used for some of the background correction techniques.

The input passed to the jet finding algorithm is given by charged particles and calorimeter clusters. The track cuts are chosen to obtain a uniform charged track distribution in the full $\eta - \phi$ plane. Additionally, only tracks from $|\eta| < 0.9$ and with $p_T > 150$ MeV/c are used in the jet finding procedure.

While *charged jets* are reconstructed using only the charged part of a jet, *full jets* also include neutral jet energy reconstructed from calorimeter clusters. The advantage of analyzing charged jets with ALICE is mainly the acceptance: as already described above, EMCal covers only a limited part in both pseudorapidity and azimuth. Since the EMCal is also sensitive to charged tracks, the energy of each cluster has to be corrected for charged energy to avoid double-counting. This is done by extrapolating tracks to the EMCal and matching them to the deposited energy. The corrected cluster energy is given by

$$E_{\text{clus}}^{\text{corr}} = E_{\text{clus}}^{\text{orig}} - \sum p^{\text{matched}}. \tag{1}$$

Eventually, only clusters with $E_{\text{clus}}^{\text{corr}} > 300$ MeV are passed to FastJet. After constituent selection, the tracks and calorimeter clusters are clustered by FastJet using the so-called p_T-recombination scheme, assuming massless jets.

In ALICE, jets have been measured with several jet resolution parameters (often called *jet radii*) for different collision systems. To avoid edge effects, only jets fully contained within the acceptance are used for further analysis.

Due to the very high multiplicities in heavy-ion collisions, there is a non-negligible contribution of jets at lower p_T made of so-called combinatorial jets mostly containing background. To suppress those jets, a bias on the leading particle's transverse momentum can be imposed. On the other hand, this biases the jets to show a harder fragmentation but, eventually, the systematic uncertainty from the fragmentation bias is smaller than that from combinatorial jets.

3.1. Background correction

In principle, every particle that does not originate from the hard parton scattering in question can be considered as background. Like the definition of a jet, also the definition of the background is not unambiguous. All presented analyses use background correction techniques but the methods differ depending on the considered collision system. While the background subtraction is a large correction to the jet momentum in Pb–Pb collisions, the background density in pp and p–Pb collisions is roughly smaller by a factor of 100 [8][9].

A typical approach to correct for background is to subtract the background on a jet-by-jet basis. The average background energy density ρ is evaluated using k_T jets (p–Pb and Pb–Pb) or by probing the background in a cone perpendicular to the jet (pp) on an event-by-event basis. Then, the transverse momentum of every jet is corrected for the background momentum density ρ depending on the jet area A by

$$p_{\text{T, jet}}^{\text{corr}} = p_{\text{T, jet}}^{\text{orig}} - \rho A. \tag{2}$$

In p–Pb and Pb–Pb collisions, region-to-region fluctuations of the background are also taken into account in an unfolding procedure. These are quantified by probing the transverse momentum in randomly distributed cones and comparing it to the average background. Quantitatively, this distribution is given by

$$\delta p_T = \sum_{\text{i}} p_{\text{T,i}} - \rho A, \quad A = \pi R^2. \tag{3}$$

3.2. Unfolding

Spectra in p–Pb and Pb–Pb collisions are corrected for background fluctuations and detector effects – e.g. from the limited single-particle tracking efficiency – in an unfolding procedure. In pp collisions, background fluctuations are neglected and only detector effects are corrected for. Depending on the analysis, Bayesian, SVD, or χ^2-unfolding or also a conceptually simpler bin-by-bin correction is used. The detector response matrix is created with a full detector simulation using PYTHIA6 [10] to generate jets and GEANT3 [11] for the particle transport through the detector.

4. Results

In the following, the results on jet production measurements are presented in terms of the nuclear modification factors and jet cross section ratios.

The nuclear modification factor shows the effect of the nuclear environment on jet production. It is defined by

$$R_{\mathrm{AA}} = \frac{\left.\frac{\mathrm{d}N}{\mathrm{d}p_{\mathrm{T}}\mathrm{d}\eta}\right|_{\mathrm{AA}}}{\langle T_{\mathrm{AA}}\rangle \cdot \left.\frac{\mathrm{d}\sigma}{\mathrm{d}p_{\mathrm{T}}\mathrm{d}\eta}\right|_{\mathrm{pp}}} \tag{4}$$

where T_{AA} is the nuclear overlap function [12], which connects the heavy-ion collisions in Pb–Pb or p–Pb to a reference measurement in pp.

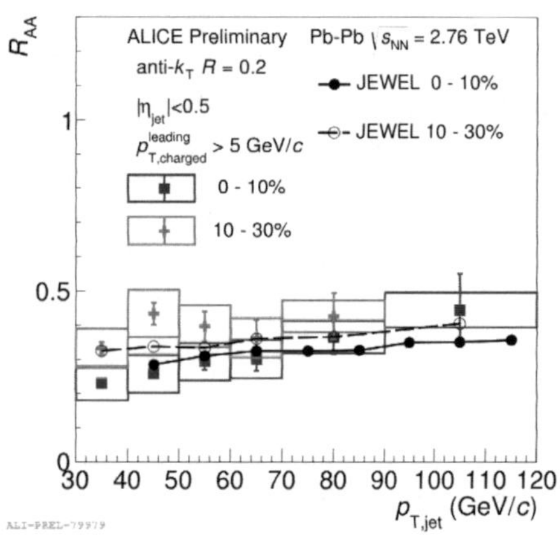

Figure 2. Nuclear modification factor for $R = 0.2$ (Pb–Pb, full jets, $\sqrt{s_{\mathrm{NN}}} = 2.76$ TeV)

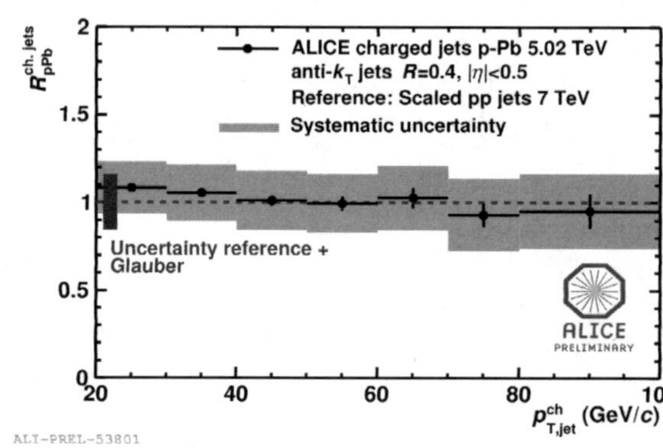

Figure 3. Nuclear modification factor for $R = 0.4$ (p–Pb, charged jets, $\sqrt{s_{\mathrm{NN}}} = 5.02$ TeV)

Preliminary ALICE measurements for full jets in Pb–Pb [13] and charged jets in p–Pb [14] are presented here, see Figs. 2 and 3, respectively. The data in Pb–Pb clearly shows jet quenching, i.e. a strong jet suppression, consistent with the charged hadron result [15]. In addition, the measurement shows that the suppression has a centrality dependence and a small transverse momentum dependence.

For p–Pb collisions (Fig. 3), no nuclear modification is observed within uncertainties. This is a strong hint that the jet quenching originates mainly from medium effects due to the formation of a quark-gluon plasma and not from cold nuclear effects due to the presence of the Pb-nucleus.

A conceptually simpler quantity is R_{CP}, the ratio of central over peripheral jet spectra scaled by the respective number of binary collisions. Like the nuclear modification factor, it is a measure

IOP Conf. Series: Journal of Physics: Conf. Series **805** (2017) 012008 doi:10.1088/1742-6596/805/1/012008

of the nuclear modification but it is simpler because no pp reference is needed. It is shown for charged jets in Pb–Pb collisions [16] in Fig. 4. Like the full jet R_{AA}, the charged jet R_{CP} shows a centrality-dependent suppression that only has a small transverse momentum dependence.

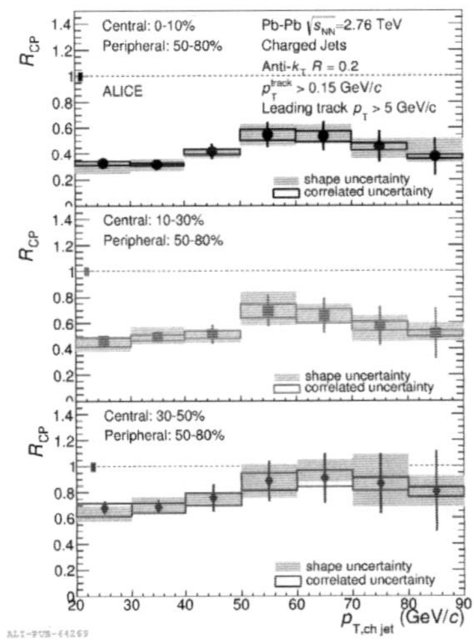

Figure 4. Central to peripheral ratio of jets for $R = 0.2$ (Pb–Pb, charged jets, $\sqrt{s_{NN}} = 2.76$ TeV)

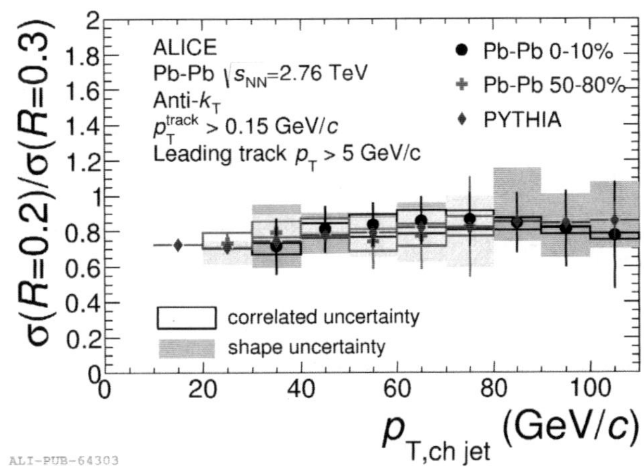

Figure 5. Jet cross section ratio compared to PYTHIA6 calculations (Pb–Pb, charged jets, $\sqrt{s_{NN}} = 2.76$ TeV)

A basic observable to study the internal radial structure of jets, i.e. jet broadening and narrowing, is the ratio of jet cross sections reconstructed with different resolution parameters. For pp collisions, the ratio has been measured for full (Fig. 6) and charged jets (Fig. 7) at different energies. The full jet analysis in pp [17] shows a good theory-data agreement when comparing to an NLO calculation with hadronization effects. The charged jet analysis in pp [18][19] shows a good agreement with PYTHIA and HERWIG [20] simulations.

For charged jets in Pb–Pb (Fig. 5), no significant medium modification of the radial jet structure is observed comparing to pp collisions. It is compatible with PYTHIA6 simulations. The same holds for p–Pb collisions: The radial structure in both, full (cf. [21]) and charged jets in p–Pb, agree well with proton-proton results (Figs. 8 and 9).

Thus, the radial jet structure is essentially unmodified by the nuclear environment and the medium in p–Pb and Pb–Pb collisions.

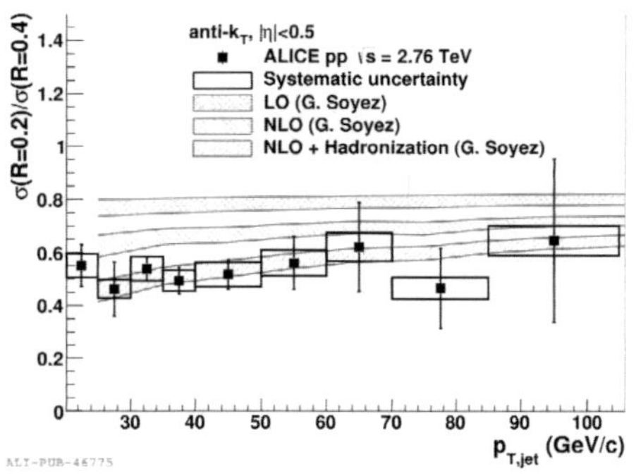

Figure 6. Jet cross section ratio (pp, full jets, $\sqrt{s} = 2.76$ TeV)

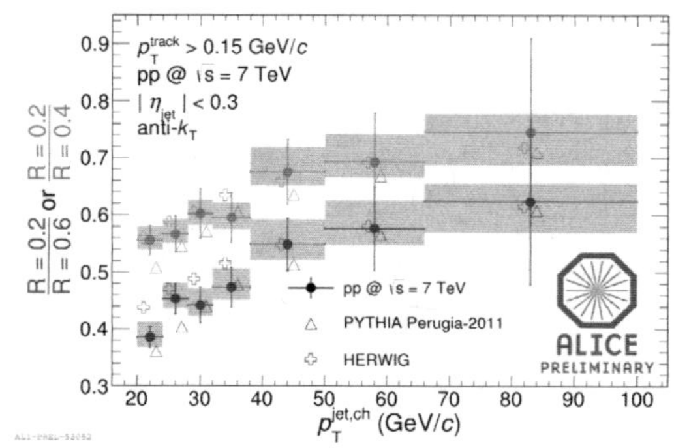

Figure 7. Jet cross section ratio (pp, charged jets, $\sqrt{s} = 7$ TeV)

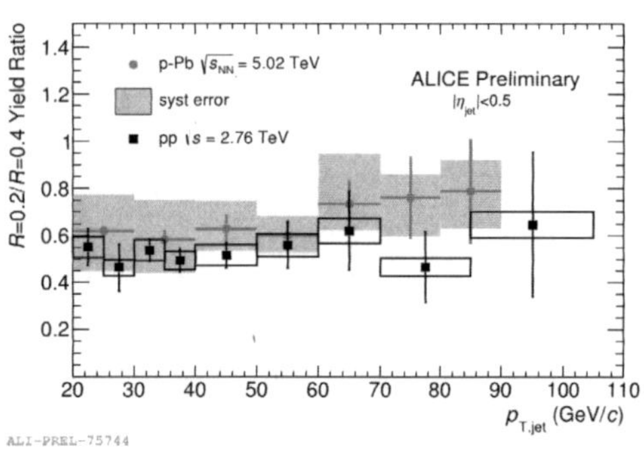

Figure 8. Jet cross section ratio in p–Pb (full jets, $\sqrt{s_{NN}} = 5.02$ TeV) compared to that in pp at $\sqrt{s} = 2.76$ TeV

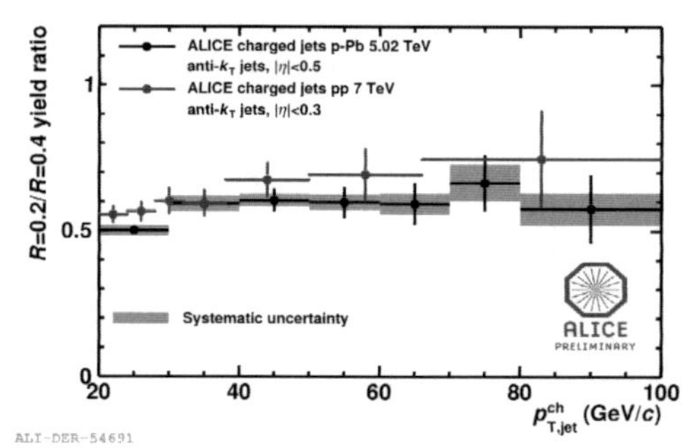

Figure 9. Jet cross section ratio in p–Pb (charged jets, $\sqrt{s_{NN}} = 5.02$ TeV) compared to that in pp at $\sqrt{s} = 7$ TeV

5. Summary

ALICE has measured charged and full jets in pp, p–Pb, and Pb–Pb collisions at several energies. In pp, the spectra and radial structure ratios for different resolution parameters show a good agreement with (next-to-)leading-order QCD calculations. The same also holds for p–Pb collisions: In p–Pb collisions, there is no sign for a nuclear modification within the uncertainties. In Pb–Pb collisions, jet quenching and its centrality dependence has been measured for full and charged jets. Neither in p–Pb nor in Pb–Pb, the jet cross section ratio shows a significant modification.

[1] Aamodt K *et al.* (ALICE Collaboration) 2008 *JINST* **3** 8002
[2] Aamodt K *et al.* (ALICE Collaboration) 2010 *JINST* **5** 3003 (*Preprint* 1001.0502)
[3] Alme J *et al.* (ALICE Collaboration) 2010 *Nucl. Instrum. Meth. A* **622** 316–367 (*Preprint* 1001.1950)
[4] Cortese P *et al.* (ALICE Collaboration) *CERN-LHCC-2008-014*
[5] Aamodt K *et al.* (ALICE Collaboration) 2010 *Phys. Rev. Lett.* **105** 252301 (*Preprint* 1011.3916)
[6] Cacciari M, Salam G P and Soyez G 2008 *JHEP* **04** 063 (*Preprint* 0802.1189)
[7] Cacciari M and Salam G P 2006 *Phys. Lett. B* **641** 57–61 (*Preprint* 0512210)
[8] Abelev B *et al.* (ALICE Collaboration) 2012 *JHEP* **03** 053 (*Preprint* 1201.2423)
[9] Cacciari M and Salam G P 2008 *Phys. Lett. B* **659** 119–126 (*Preprint* 0707.1378)
[10] Sjoestrand T, Mrenna S and Skands P 2006 *JHEP* **05** 026 (*Preprint* hep-ph/0603175)
[11] Brun R, Carminati F and Giani S 1994 *CERN Program Library Long Writeup CERN-W-5013*

IOP Conf. Series: Journal of Physics: Conf. Series **805** (2017) 012008 doi:10.1088/1742-6596/805/1/012008

[12] Abelev B *et al.* (ALICE Collaboration) 2013 *Phys. Rev. C* **88** 44909 (*Preprint* 1301.4361)
[13] Aiola S (ALICE Collaboration) 2014 *QM2014 proceedings, to be published* (*Preprint* 1408.0479)
[14] Haake R (ALICE Collaboration) 2013 *PoS EPS-HEP2013 (proceedings)* 176 (*Preprint* 1310.3612)
[15] Aamodt K *et al.* (ALICE Collaboration) 2011 *Phys. Lett. B* **696** 30–39 (*Preprint* 1012.1004)
[16] Abelev B *et al.* (ALICE Collaboration) 2014 *JHEP* **03** 013 (*Preprint* 1311.0633)
[17] Abelev B *et al.* (ALICE Collaboration) 2013 *Phys. Lett. B* **722** 262–272 (*Preprint* 1301.3475)
[18] Vajzer M (ALICE Collaboration) 2013 *PoS EPS-HEP2013 (proceedings)* 464 (*Preprint* 1311.0148)
[19] Abelev B *et al.* (ALICE Collaboration) 2014 *to be published*
[20] Corcella G, Knowles I G, Marchesini G, Moretti S, Odagiri K, Richardson P, Seymour M H and Webber B R 2001 *JHEP* **2001** 010010 ISSN 1029-8479
[21] Connors M (ALICE Collaboration) 2014 *LHCP proceedings, to be published* (*Preprint* 1409.4655)

IOP Conf. Series: Journal of Physics: Conf. Series **805** (2017) 012009 doi:10.1088/1742-6596/805/1/012009

Production of strange particles in charged jets in p–Pb and Pb–Pb collisions measured with ALICE at the LHC

Alice Zimmermann[1] **(for the ALICE Collaboration)**

1. Physikalisches Institut, Ruprecht-Karls-Universität Heidelberg, Im Neuenheimer Feld 226, 69120 Heidelberg, Germany

alice.zimmermann@cern.ch

Abstract. Studies of jet production can provide information about the properties of the hot and dense strongly interacting matter created in ultra-relativistic heavy-ion collisions. Specifically, measurement of strange particles in jets may clarify the role of fragmentation processes in the anomalous baryon to meson ratio at intermediate particle p_T that was observed in lead-lead (Pb–Pb) and, to a lesser extent, in proton-lead (p–Pb) collisions.

In this contribution, measurements are presented of the p_T spectra of $\Lambda(\overline{\Lambda})$ baryons and K^0_S mesons produced in association with charged jets in Pb–Pb collisions at $\sqrt{s_{NN}} = 2.76$ TeV and p–Pb collisions at $\sqrt{s_{NN}} = 5.02$ TeV. The analysis is based on data recorded by ALICE at the LHC, exploiting its excellent particle identification capabilities. The baryon/meson ratios of strange particles associated with jets are studied for different event activities in p-Pb collisions and are restricted to central events in Pb-Pb collisions. A comparison is shown to the ratios obtained for inclusive particle production and for particles stemming from the underlying event as well as to PYTHIA proton-proton (pp) simulations.

1. Introduction

The first measurements of the baryon-meson ratio in heavy-ion collisions at the Relativistic Heavy Ion Collider (RHIC) showed an enhanced baryon/meson production at intermediate transverse momentum ($p_T = 3$ GeV/c) [1, 2] relative to pp collisions. Recent results at LHC energy corroborated this observation [3, 4]. There are several scenarios proposed to explain this, e.g. collective effects and string fragmentation in a hydrodynamically expanding environment [5]. These collective phenomena [6] are a characteristic feature of the Underlying Event (UE) in Pb–Pb collisions, that are independent of the fragmentation process but could have a possible impact on the jet fragments. Interactions between partons, stemming from the fragmentation shower and the Quark-Gluon Plasma, could change the jet pattern, which was already observed as quenching of jets [7]. Some other models consider alternative hadronisation mechanisms, e.g. recombination at low p_T [8], which is a soft process that could favour baryon over meson production. Particles with a momentum larger than p_T = 4-6 GeV/c would, on the other hand, be produced in hard processes via fragmentation, that does not lead to an enhanced production of baryons (compared to production in vacuum). The study of identified particle yields and ratios like $\Lambda(\overline{\Lambda})$ and K^0_S in inclusive production and comparison to production in jets will allow to trace

IOP Conf. Series: Journal of Physics: Conf. Series **805** (2017) 012009 doi:10.1088/1742-6596/805/1/012009

V0 UE estimation method	V0 sample
No-jet events	in events without any selected jets (reconstructed jets fulfilling all further requirements)
Outside cones	outside of 2R (R = 0.2,0.3)
Random cones	in cones with randomly chosen axis (η,ϕ) and no overlap with selected jets
Median-cluster cones	uses median k_t cluster (similar to k_t alg. for average background estimation)
Perpendicular cones	in cone perpendicular (in ϕ) to jet axis

Table 1. Methods to estimate the V0 particle contribution from the UE

back possible differences in the hadronisation process and its p_T dependence. Since collective features were proposed to also occur in p–Pb collisions, one can compare among the different collision systems (pp, p–Pb and Pb–Pb), to study the onset of this baryon enhancement, that has already been observed for inclusive production in p–Pb collisions [9]. For the analyses presented in this contribution, the p_T spectra of strange particles associated with jets are studied for events with different event multiplicities in p–Pb collisions and for the 10% most central events in Pb–Pb collisions.

2. Experimental setup and analysis strategy

The jet reconstruction is done with the $anti - k_t$ jet finder using charged tracks with $p_T > 150$ MeV/c and requiring a leading particle in each jet to have at minimum $p_T = 5$ GeV/c. In the following, jets fulfilling all these requirements are referred to as 'selected' jets. The jet energy is corrected on an event-by-event basis for the contribution of the average charged UE [10].

Both p–Pb and Pb–Pb analyses use the decay topology of the strange neutral particles (V0s) for their reconstruction. The decay channels $K^0_S \rightarrow \pi^+ + \pi^-$ [11] and $\Lambda \rightarrow p + \pi^-$ ($\overline{\Lambda} \rightarrow \overline{p} + \pi^+$) [11] are used. By recombining the charged tracks of opposite curvature, measured in the Inner Tracking System (ITS) and the Time Projection Chamber (TPC), together with a particle mass hypothesis the invariant mass of the V0 candidates is calculated. A set of cuts applied on various observables of the topology of the decay already rejects a sizeable amount of combinatorial background. The particle signal is extracted after a background fit and by applying a bin-counting procedure on the invariant mass distributions in different intervals of p_T^{V0} and $p^{ch}_{T,\,jet}$. The V0 candidate is associated to the jet cone if the distance, which is calculated in azimuthal angle (ϕ) and in pseudo-rapidity (η), to the jet axis is smaller than the jet resolution parameter R:

$$\sqrt{(\phi_{V0} - \phi_{jet})^2 + (\eta_{V0} - \eta_{jet})^2} < R \tag{1}$$

The V0 acceptance is chosen to $|\eta^{V0}| < 0.7$, the acceptance for the jets is $|\eta^{jet}| < |\eta^{V0}| - R = 0.5(0.4)$ for R = 0.2(0.3).

Figure 1 shows the uncorrected K^0_S and Λ p_T spectra for jets with R = 0.3 and for two different jet p_T intervals ($p^{ch}_{T,\,jet} > 10$ GeV/c, > 20 GeV/c) in Pb–Pb collisions. The spectra are scaled for better visibility.

Several corrections have to be applied to the raw yields. First, one has to correct for the V0 particle reconstruction efficiency as functions of p_T and η, by using Monte-Carlo simulations. The efficiency is calculated separately for V0s in the inclusive event, the jet cone (JC) and in the UE. Since V0s from the UE contribute quite substantially to the V0 spectra in the JC, especially at low p_T, this has to be accurately subtracted. For this UE V0 correction, we developed several methods (see table 1) to measure the raw V0 p_T spectrum in the UE as shown in Figure 2. The default method are inclusive V0s taken from events, in which no reconstructed jet is selected; this is the so-called "No-jet events" method. To assign a systematic uncertainty to this correction, 4 alternative methods are applied. Compared to the default method (see ratio in Figure 2, bottom), they show a deviation of around 15% for $p_T^{V0} < 4$ GeV/c. The order of the different correction steps is as follows: After the correction for the V0 reconstruction efficiency is applied,

IOP Conf. Series: Journal of Physics: Conf. Series **805** (2017) 012009 doi:10.1088/1742-6596/805/1/012009

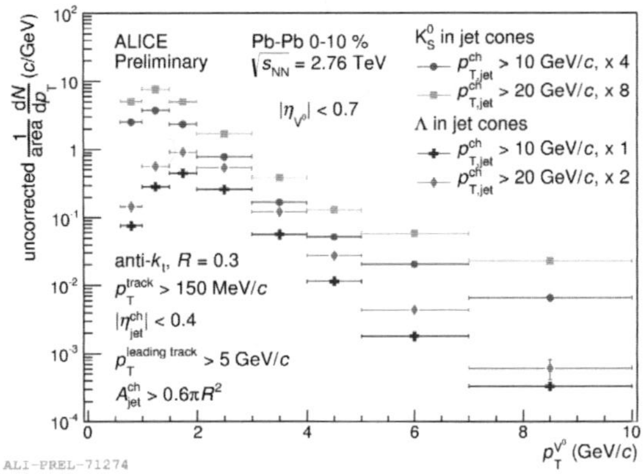

Figure 1. Uncorrected K^0_S and Λ p_T spectra for R = 0.3 jets in Pb–Pb collisions at 2.76 TeV and in the 10% highest multiplicity class. The jets are reconstructed for two different jet p_T intervals ($p^{ch}_{T,\,jet} > 10$ GeV/c, > 20 GeV/c) with the $anti-k_t$ jet finder using charged tracks with minimum $p_T > 150$ MeV/c.

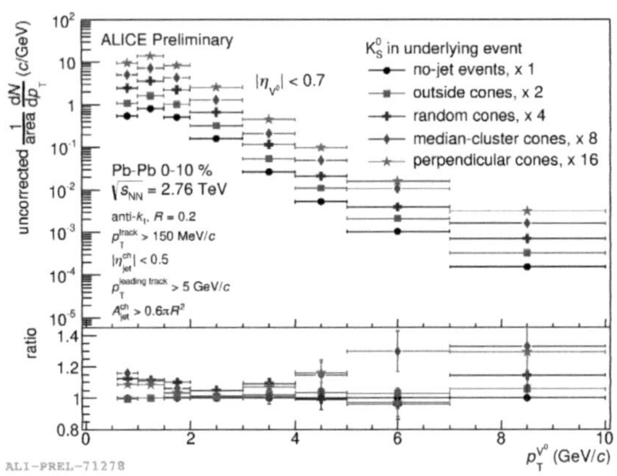

Figure 2. (top) Uncorrected K^0_S p_T spectra in UE in Pb–Pb collisions at $\sqrt{s_{NN}}$ = 2.76 TeV and in the 10% most central events. The different used methods serve to estimate systematic uncertainty of UE subtraction. (bottom) Each method divided by default (No-jet method), the ratios show a deviation of around 15% for $p_T^{V0} < 4$ GeV/c.

the UE V0 contribution (using the "no-jets" method) and the feed-down (FD) are subtracted from the V0 p_T spectrum in the JC. To account for the sizable FD into Λ mostly stemming from $\Xi^{0,-} \rightarrow \Lambda$ and since there is no direct $\Xi^{0,-}$ measurement in jets so far, two extreme scenarios are considered for the V0s in the JC, namely the FD contribution from the inclusive particle analysis [3] and the correction derived from PYTHIA pp simulations. The difference is used to assign a systematic uncertainty.

3. Results

Figure 3 shows the Λ/K^0_S ratio in jets in p–Pb events at 5.02 TeV and in the 10% highest multiplicity class. It is compared to the inclusive Λ/K^0_S ratio (solid lines) in p–Pb collisions and to the ratio predicted by PYTHIA 8 4C simulations at 5.02 TeV/c (dotted lines, showing spread for the used values of R). The measured ratio in jets shows no visible enhancement at intermediate p_T and is far below the inclusive ratio. Within the systematic uncertainties, the ratio in jets is in agreement with the PYTHIA simulations. Furthermore there is no significant dependence on the $p^{ch}_{T,\,jet}$ interval or on R visible.

4. Conclusions

The observed Λ/K^0_S ratio in jets in p–Pb collisions is significantly smaller than the inclusive measurement and compatible with the predictions of a PYTHIA jet simulation. This points to a

IOP Conf. Series: Journal of Physics: Conf. Series **805** (2017) 012009 doi:10.1088/1742-6596/805/1/012009

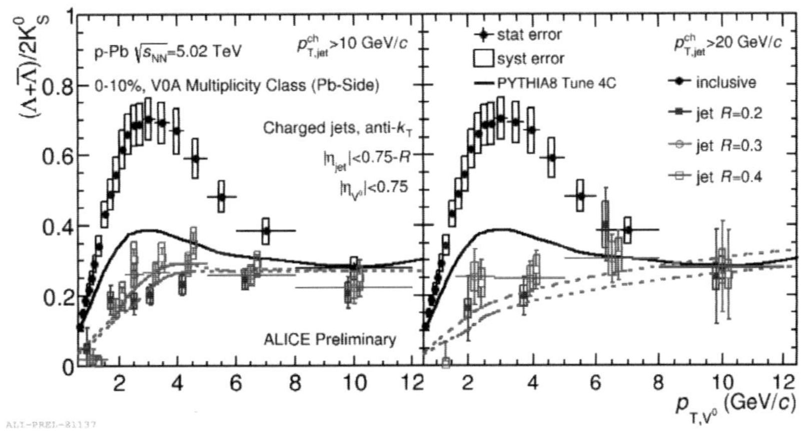

Figure 3. $\Lambda/\mathrm{K}^0_\mathrm{S}$ ratio in jets for $p^{\mathrm{ch}}_{\mathrm{T,\,jet}} > 10$ GeV/c (left) and $p^{\mathrm{ch}}_{\mathrm{T,\,jet}} > 20$ GeV/c (right) measured in p–Pb collisions at $\sqrt{s_{\mathrm{NN}}} = 5.02$ TeV for events in the 0-10% V0A (determined by the VZERO-A detector) multiplicity class [12, 13]. Results of the ratio in jets are compared to the inclusive particle ratio (solid lines) and the expectation from a PYTHIA 8 4C simulation at 5.02 TeV/c (dotted lines, showing spread for the used values of R).

scenario in which the hadronisation of soft particles is disentangled from the hard fragmentation processes. Therefore the baryon enhancement seen in inclusive production (see black markers in Figure 3) at intermediate particle p_T in p–Pb collisions seems to be a feature of the UE and not from fragmentation. Furthermore, the uncorrected spectra for K^0_S and Λ particles in the JC and in UE in 10% most central events in Pb–Pb collisions have been presented. The $\Lambda/\mathrm{K}^0_\mathrm{S}$ ratio in jets in Pb–Pb collisions will be reported soon.

References

[1] Agakishiev G *et al.* (STAR Collaboration) 2012 *Phys. Rev. Lett.* **108** 072301 (`arXiv:nucl-ex/1107.2955`)
[2] Adams J *et al.* (STAR Collaboration) 2006 (`arXiv:nucl-ex/0601042v1`)
[3] Abelev B *et al.* (ALICE Collaboration) 2013 *Phys. Rev. Lett.* **111** 222301 (`arXiv:nucl-ex/1307.5530`)
[4] Abelev B *et al.* (ALICE Collaboration) 2014 *Phys. Lett.* **B736** 196207 (`arXiv:nucl-ex/1401.1250v4`)
[5] Werner K 2012 *Phys. Rev. Lett.* **109** 102301 (`arxiv:nucl-th/1204.1394`)
[6] Aamodt K *et al.* (ALICE Collaboration) 2010 *Phys. Rev. Lett.* **105** (`arXiv:nucl-ex/1405.4632`)
[7] Milhano J G 2013 *Nucl. Phys.* **A904-905** 122c–129c (`arxiv:hep-ex/1211.2657v1`)
[8] Fries R, Müller B, Nonaka C and Bass S 2003 *Phys. Rev. Lett.* **90** 202303 (`arxiv:nucl-th/0301087`)
[9] Abelev B *et al.* (ALICE Collaboration) 2014 *Phys. Lett.* **B728** 25 (`arxiv:nucl-ex/1307.6796`)
[10] Cacciari M and Salam G P 2008 *Phys. Lett.* **B659** 119126 (`arXiv:hep-ex/0707.1378`)
[11] Olive K A *et al.* (Particle Data Group) 2014 *Chin. Phys.* **C38** 090001
[12] Aamodt K *et al.* (ALICE Collaboration) 2008 *JINST* **3** S08002 (`doi:10.1088/1748-0221/3/08/S08002`)
[13] Abelev B *et al.* (ALICE Collaboration) 2014 *Int. J. Mod. Phys.* **A** 1430044 (`arxiv:nucl-ex/1402.4476`)

A 'soft + hard' model for heavy-ion collisions

K Urmossy [a], **G G Barnaföldi** [a], **Sz Harangozó** [a,b], **T S Biró** [a], **and Z Xu** [c]

[a] Wigner Research Center for Physics of the Hungarian Academy of Sciences
29–33 Konkoly–Thege Miklós Str., H-1121 Budapest, Hungary

[b] Eötvös University, 1/a Pázmány Péter Walkway, H-1117 Budapest, Hungary

[c] Physics Department, Brookhaven National Laboratory, Upton, NY 11973, USA

E-mail: karoly.uermoessy@cern.ch

Abstract. We describe the spectra of neutral pions stemming from AuAu collisions at $\sqrt{s}$ = 200 AGeV in a 'soft + hard' model. The model is based on the assumptions that hadrons stemming from hard processes are described via perturbative quantum chromodynamics improved parton model calculations, while those stemming from the Quark-Gluon Plasma (we refer to as soft yields) can be described in a super-statistical model induced by multiplicity fluctuations. The obtained dependence of the parameters of the model on the event centrality is compared to what is observed in PbPb collisions at $\sqrt{s} = 2.76$ ATeV.

1. Introduction

Among others, the phenomenon of jet-quenching supports that in ultra-relativistic heavy-ion collisions (AA) a new hadron source the Quark-Gluon Plasma (QGP) emerges besides the hard processes present in proton-proton (pp) and proton-nucleus (pA) collisions. We refer to hadrons stemming from these two types of sources as 'soft' and 'hard' yields. While perturbative quantum-chromodynamics improved parton model (pQCD) calculations [1, 2, 3] describe the spectra of hadrons stemming from hard processes (pp, pA collisions and the high transverse momentum $p_T \gtrsim 6$ GeV/c part of the spectra in AA collisions), pQCD methods fail to describe the $p_T \lesssim 6$ GeV/c part of hadron spectra in AA collisions.

Disentangling soft and hard yields is a cumbersome problem for there may be low-p_T hadrons stemming from hard processes as well as moderately high-p_T ones created by soft processes [34]. Another difficulty is that due to the break-down of perturbation theory at low p_T, we might not trust pQCD results at low momenta. In this paper, we demonstrate two possible ways of obtaining the low-p_T part of the hard hadron yields: (i), we simply take the pQCD results; (ii), we use a function fitted to the high-p_T part of the pQCD spectra to make an extrapolation to low p_T. In the second case, we use a cut-power law function for the extrapolation, as such function turned out to provide an adequate description for hard hadron yields produced in pp collisions [4]–[17].

Once hard hadron yields (obtained in the (i) and (ii) scenarios) are subtracted from the measured transverse hadron spectra, we describe the remaining soft yields in the often used super-statistical framework [12, 19, 20, 21, 28, 31, 33, 40, 41]. In this paper, we use the interpretation in which, the cut-power law shape of the soft hadron distribution originates from multiplicity fluctuations rather than from non-extensive entropy or transport [29, 32].

IOP Conf. Series: Journal of Physics: Conf. Series **805** (2017) 012010 doi:10.1088/1742-6596/805/1/012010

The next section contains fits to neutral pion spectra measured in various centrality AuAu collisions at $\sqrt{s} = 200$ AGeV by the PHENIX Collaboration [42]. To draw conclusions, we also discuss results obtained examining ALICE and CMS measurements [34].

We note that similar soft and hard type separation of hadron yields has been suggested in models [22, 23, 24] in which, spectra measured in *pp* collisions have been used as hard yields, and it has been conjectured that hard yields are suppressed at low p_T. Similar analysis of the soft part of hadron spectra without the subtraction of hard yields are also presented in the literature [18]–[31].

2. Transverse spectrum in the 'soft + hard' model

As argued in Sec. 1 and in [34], we make out the transverse spectrum of hadrons from the sum of hard and soft yields

$$\left.\frac{dN}{2\pi p_T dp_T dy}\right|_{y=0} = f_{hard} + f_{soft} \,. \tag{1}$$

As can be seen in the top-right panel of Fig. 1, pQCD results describe transverse spectra of neutral pions stemming from various centrality AuAu collisions at $\sqrt{s} = 200$ AGeV for $p_T \geq 6$ GeV/c. In the following, we examine two scenarios:

i, We use pQCD results as hard yields for $p_T \leq 6$ GeV/c too, and conjecture that the soft yields can be described by a cut-power law function.

ii, We fit pQCD results for $p_T \geq 6$ GeV/c with one cut-power law. We use this fitted function to extrapolate the hard yields to low p_T. Finally, we use a cut-power law with different parameters to fit the soft part of the spectra.

In this approach, both the $p_T \leq 6$ GeV/c and the $p_T \geq 6$ GeV/c regions contain soft as well as hard yields. The cut-power law functions are

$$f_i = A_i \left[1 + \frac{(q_i - 1)}{T_i}[\gamma_i(m_T - v_i p_T) - m]\right]^{-1/(q_i-1)} \tag{2}$$

where $i =$ soft or hard, $m = m_{\pi^0}$. Thus, q_i, T_i, v_i and A_i are the fit parameters of the model. The transverse mass is $m_T = \sqrt{p_T^2 + m^2}$, and v_i is the isotropic part of the transverse flow with $\gamma_i = 1/\sqrt{1 - v_i^2}$ (see [34] for a detailed discussion). These yields have maxima at $p_T^{max} = \gamma_i \, m \, v_i$. As long as these maxima are below the measurement range, which is the case in this analysis, the transverse flow v_i cannot be determined accurately. As in this paper we analyse neutral pions, the arguments of f_i-s may be approximated by $[\gamma_i(m_T - v_i p_T) - m]/T_i \approx p_T/T_i^{Dopp}$ with the Doppler-shifted parameters

$$T_i^{Dopp} = T_i \sqrt{\frac{1 + v_i}{1 - v_i}} \,. \tag{3}$$

Results in the (*i*) scenario

Fit results of Eqs. (1)–(2) in the two scenarios explained above, are shown in Fig. 1. The dependence of the obtained q and T^{Dopp} parameters of the soft and hard yields on the event centrality (number of participating nucleons N_{part}) can be fitted by (see Fig. 2)

$$\begin{aligned} q_i &= q_{2,i} + \mu_i \ln(N_{part}/2) \,, \\ T_i^{Dopp} &= T_{1,i} + \tau_i \ln(N_{part}) \end{aligned} \tag{4}$$

with fit parameters $q_{2,i}, T_{1,i}, \mu_i$ and τ_i.Though the actual value of the transverse flow velocity cannot be determined in this model from the spectra of charged hadrons, it may be guessed

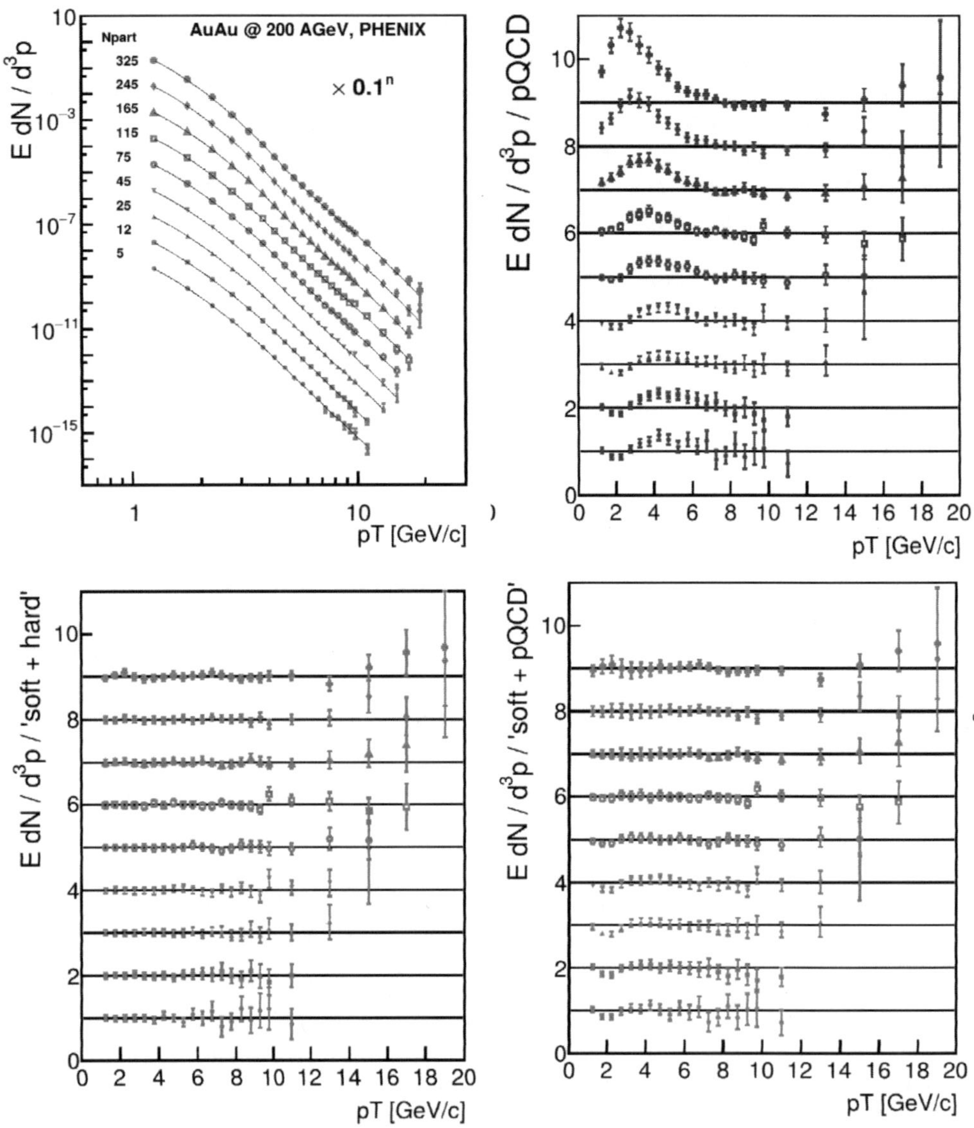

Figure 1. Top-left, transverse spectra of neutral pions stemming from various centrality AuAu collisions at $\sqrt{s} = 200$ AGeV [42]. Topmost graphs are central events, bottommost graphs are peripheral events in all panels. Solid curves: fits of Eqs. (1)–(2), in which, both the soft and hard yields are fitted cut-power law functions as described in the (i) scenario. **Bottom-left,** Ratio of measured data and fits in the (i) scenario. **Top-right,** ratio of measured yields and pQCD results. **Bottom-right,** ratio of measured yields and model results, in which, the hard yields are pQCD results and the soft yields are fitted cut-power law functions as described in the (ii) scenario.

IOP Conf. Series: Journal of Physics: Conf. Series **805** (2017) 012010 doi:10.1088/1742-6596/805/1/012010

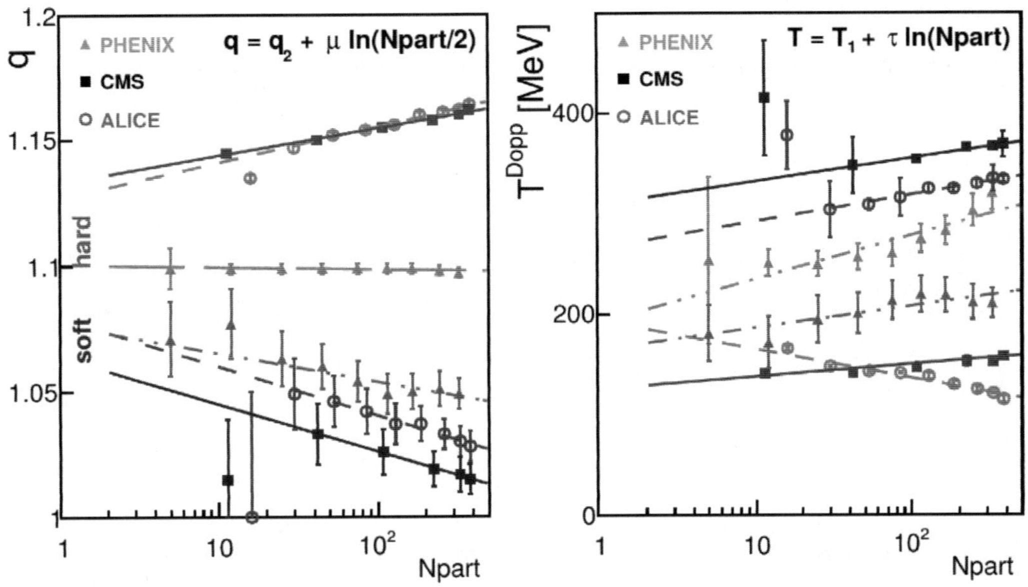

Figure 2. Fitted parameters of Eqs. (1) – (2) in the (*i*) scenario to PHENIX data on the spectra of π^0-s stemming from AuAu collisions at $\sqrt{s} = 200$ AGeV (see Fig. 1). For comparison, results of fits to charged hadron spectra in PbPb collisions at $\sqrt{s} = 2.76$ ATeV measured by the CMS and ALICE collaborations are also shown (see [34]). In both panels, data points belonging to soft hadron yields are colored in shades of blue, whereas those belonging to hard yields are reddish. Curves are fits of Eq. (4).

using the value of the QGP-hadronic matter transition temperature obtained from lattice-QCD calculations. As the values of fitted T_{soft}^{Dopp} scatter around ~200 MeV at RHIC (and ~340 MeV at LHC), in case of a flow velocity of $v_{soft} \approx 0.5$ at RHIC (and $v_{soft} \approx 0.6$ at LHC), the real T_{soft} values would scatter around 170 MeV, which is close to the lattice result obtained e.g. in [43].

q_{soft} and T_{soft} may have statistical meaning. If in each nucleus-nucleus collision, hadrons stemming from soft processes have either Boltzmann–Gibbs or micro-canonical distribution, however, hadron multiplicity fluctuates according to the negative-binomial distribution, then the cut-power law function Eq. (2) emerges as the average hadron spectrum [12, 19, 20, 21, 28, 31, 33, 40, 41]. If $q \to 1$, Eq. (2) tends to the Boltzmann distribution, while the multiplicity distribution tends to the Poissonian distribution. Such tendency can be seen in the left panel of Fig. 2. We may interpret this phenomenon as the soft yields are closer to the thermal, if either N_{part} or $\sqrt{s}$ grows. As can be seen in the top panels of Fig. 1 and right panel of Fig. 2, the more central the collision, the bigger and hotter the soft yield.

The behaviour of the parameters of the hard yields have to be determined by scale evolution equations of QCD in yet unexamined ways. Nevertheless, it can be seen that as $\sqrt{s}$ increases, q_{hard} increases as well, while the dimensionless parameter $T_{hard}/\sqrt{s}$ appearing in the cut-power-law-type fragmentation functions introduced in [3], decreases. These trends seen in Fig. 2 are similar to what has been found in [3].

	$q_{2,soft}$	$q_{2,hard}$	μ_{soft}	μ_{hard}
CMS	1.058 ± 0.025	1.136 ± 0.001	-0.008 ± 0.005	0.005 ± 0.0003
ALICE	1.074 ± 0.018	1.131 ± 0.002	-0.009 ± 0.004	0.006 ± 0.0006
PHENIX	1.073 ± 0.016	1.100 ± 0.002	-0.005 ± 0.004	0.000 ± 0.0006
	T_1^{soft} [MeV]	T_1^{hard} [MeV]	τ_{soft} [MeV]	τ_{hard} [MeV]
CMS	310 ± 20	126 ± 5	9.9 ± 3.7	5.3 ± 0.8
ALICE	266 ± 16	194 ± 2	11.5 ± 2.9	-12.5 ± 0.5
PHENIX	165 ± 26	192 ± 20	9.3 ± 5.5	18.7 ± 4.6

Table 1. Parameters of Eq. (4) obtained from fits shown in Fig. 2.

Results in the (*ii*) scenario

In the second scenario, the hard yields were taken directly from the pQCD calculations, and the soft yields were fitted by the cut-power law function Eq. (2) (see right panels of Fig. 1). The obtained soft distributions were practically exponentials, thus we may conclude that if pQCD results [1, 2] are trustworthy at $p_T \leq 6$ GeV/c, the soft yields are likely to be thermal. This scenario has been thoroughly discussed in the literature [35]–[39].

Acknowledgement

This work was supported by Hungarian OTKA grants K104260, NK106119, and NIH TET 12 CN-1-2012-0016. Author GGB also thanks the János Bolyai Research Scholarship of the Hungarian Academy of Sciences.

References

[1] Y. Zhang et.al., *Phys. Rev. C*, **65**, 034903 (2002).
[2] G. G. Barnafoldi, J. Barrette, M. Gyulassy et.al., *Phys. Rev. C*, **85**, 024903 (2012)
[3] G. G. Barnaföldi, G. Kalmár, K. Urmossy et.al, *Gribov 80 - Memorial Volume*, (ISBN: 978-981-4350-18-1), p. 357., World Scientific, Singapore, (2011)
[4] Cheuk-Yin Wong, Grzegorz Wilk, Conference: C13-05-30, (2013)
[5] Cheuk-Yin Wong, Grzegorz Wilk, *Phys. Rev. D*, **87**, 114007 (2013)
[6] The PHENIX Collab., Phys. Rev. D, 83, 052004 (2011)
[7] Cheuk-Yin Wong, G. Wilk, *Acta Physica Polonica B*, **42**, 2047, (2012)
[8] J. Cleymans, Proc. of: 1st *International Conference on New Frontiers in Physics ICFP*, (2012)
[9] J. Cleymans, D. Worku, *J. Phys. G: Nucl. Part. Phys.*, **39**, 025006, (2012)
[10] J. Cleymans, D. Worku, *Eur. Phys. J. A*, **48**, 11, 160, (2012)
[11] G. G. Barnaföldi et al., *J. Phys. Conf. Ser.*, **270**, 012008, (2011)
[12] K. Urmossy, arXiv:1212.0260
[13] T. Wibig, I. Kurp, *JHEP*, **0312**, 039, (2003);
[14] T. Wibig, *J. Phys. G: Nucl. Part. Phys.*, **37**, 115009, (2010)
[15] Khandai, *Int. J. Mod. Phys. A*, **28** (2013) 1350066,
[16] M. Praszalowicz, *Physics Letters B*, **727**, (2013) 461
[17] T. Wibig, arXiv:1304.0655
[18] M. Rybczynski, Z. Wlodarczyk, G. Wilk, *J. Phys. G: Nucl. Part. Phys.*, **39**, 095004, (2012)
[19] G. Wilk, Z. Wlodarczyk, arXiv:1203.4452 (2012)
[20] G. Wilk, Z. Wlodarczyk, *Cent. Eur. J. Phys.*, **10**, Issue 3, pp.568-575, (2011)
[21] G. Wilk, Z. Wlodarczyk, *J. Phys. G*, **38**, 065101, (2011)
[22] Z. Tang, L. Yi, L. Ruan, et al., arXiv:1101.1912, (2011)
[23] Z. Tang, Y. Xu, L. Ruan et.al., *Phys. Rev. C:* **79**, 051901 (R), (2009)
[24] M. Shao, L. Yi, Z. Tang, et al., *J. Phys. G:* **37**, 085104, (2010)

[25] D. D. Chinellato, J. Takahashi, I. Bediaga, *J. Phys. G:* **37**, 094042, (2010)
[26] K. Urmossy, T. S. Biro, *Phys. Lett. B:* **689**, 14-17, (2010)
[27] B. De, G. Sau, S. K. Biswas et.al., *J. Mod. Phys. A:* **25**, 1239-1251, (2010)
[28] G. Wilk, Z. Wlodarczyk, *Eur. Phys. J. A*, **40**, 299-312, (2009)
[29] T. S. Biro (ed.) *et al.*, *Eur. Phys. J. A*: **40** 255-344, (2009)
[30] J. Cleymans, G. Hamar, P. Levai et.al., *J. Phys. G:* **36**, 064 018, (2009)
[31] M. Biyajima, T. Mizoguchi, N. Nakajima et.al., *Eur. Phys. J. C:* **48**, 597-603, (2006)
[32] T. S. Biro, A. Jakovac, *Phys. Rev. Lett.*, **94** (2005) 132302
[33] T. S. Biro, P. Van, G. G. Barnafoldi, K. Urmossy, *Entropy* **16** (2014) 6497-6514
[34] K. Urmossy, T. S. Biró, G. G. Barnaföldi, Z. Xu, arXiv:1405.3963
[35] R. Hagedorn, *Suppl. Nuovo Cim.*, **3**, 147, (1965);
[36] M. Szczekowski, G. Wilk, *Phys. Lett. B*, **374**, 225, (1996)
[37] K. Redlich, A. Andronic, F. Beutler et.al., *J. Phys. G*, **36**, 064021, (2009)
[38] F. Becattini, G. Passaleva, *Eur. Phys. J. C*, **23**, 551-583, (2002)
[39] F. M. Liu, K. Werner, J. Aichelin et.al., *J. Phys. G*, **30**, 589-594, (2004)
[40] K. Urmossy, G. G. Barnaföldi, T. S. Biró, *Phys. Lett. B*, **718**, 125-129, (2012),
[41] K. Urmossy, G. G. Barnaföldi, T. S. Biró, *Phys. Lett. B*, **701**, 111-116, (2011),
[42] A. Adare, et al, PHENIX Collaboration, *Phys. Rev. Lett.*, **101** 232301, (2008)
[43] S. Borsanyi, Z. Fodor, C. Hoelbling et.al., *Phys. Lett. B*, **370**, (2014) 99-104

IOP Conf. Series: Journal of Physics: Conf. Series **805** (2017) 012011 doi:10.1088/1742-6596/805/1/012011

Overview of ALICE results on azimuthal correlations using neutral- and heavy-flavor triggers

Sona Pochybova for the ALICE Collaboration
MTA Wigner RCP, 29 – 33 Konkoly-Thége Miklós street, 1121 Budapest, Hungary
E-mail: sona.pochybova@cern.ch

Abstract. The ALICE detector is dedicated to studying the properties of hot and dense matter created in heavy-ion collisions. Among the probes used to investigate these properties are high-momentum particles, which originate in hard-scatterings occurring before the fireball creation. The fragments of hard scatterings interact with the hot and dense matter and via this interaction their spectra and azimuthal distributions are modified. This is probed by the measurement of the nuclear modification factor, where the p_T spectra obtained in Pb–Pb collisions are compared to a pp baseline. A strong suppression of charged hadrons as well as neutral- and heavy-flavor mesons was observed at $p_T > 4$ GeV/c. Azimuthal correlations, using high-momentum ($p_T > 4$ GeV/c) hadrons as triggers, can provide further insight into how the presence of the medium modifies the final kinematic distributions of the particles. Comparison with theoretical models can be used to test their predictions about the properties of the medium. We give an overview of ALICE azimuthal-correlation measurements of neutral- and heavy-flavor mesons with charged hadrons in pp collisions at $\sqrt{s} = 7$ TeV and Pb–Pb collisions at $\sqrt{s_{NN}} = 2.76$ TeV. We also present a measurement of the π^0 correlation with jets in pp collisions at $\sqrt{s} = 7$ TeV.

1. Introduction

In the high-energy heavy-ion collisions the creation of a new state of matter is expected. This matter is the hot and dense "primordial soup" called the *Quark-Gluon Plasma (QGP)* [1]. The ALICE (A Large Ion Collider Experiment) collaboration at the LHC is dedicated to the study of the properties of this new state of matter [2]. To do this, a set of probes are exploited. Among these are the high-momentum particles that were created from the fragments of high-momentum partons produced in the early stages of the collision. These partons may travel through the fireball probing its evolution. Partons lose energy via medium-induced gluon radiation [3, 4] or via elastic collisions with the medium constituents [5, 6]. Once the fireball freezes out, these partons hadronize into high-momentum particles that carry the information about the parton-medium interaction. We can look at how the properties of these high-momentum particles change with respect to simpler systems as pp and p–Pb collisions and obtain information about the properties of the matter they emerged from and test our expectations.

In this work, we focus on the measurements using high-momentum π^0 and D mesons as trigger particles. High-momentum neutral pions are experimentally identified via the 2-photon decay channel in a wide range of momentum and can be used to scan different regions. The D mesons consist of charm quarks, that are created in the initial stages of the collision. Due to their large masses, they are predominantly produced in hard scatterings. The energy loss these heavy quarks suffer, is expected to differ from light quarks and gluons. The gluon radiation at small angles with respect to the quark direction is suppressed in heavy-quark fragmentation, which produces a *"dead-cone"* around the original parton. This implies a smaller energy loss for heavier quarks. The dead-cone effect is also present in vacuum [7].

One of the observables sensitive to the parton-medium interaction is the *Nuclear modification factor*:

$$R_{\mathrm{AA}}(p_{\mathrm{T}}) = \frac{1}{\langle T_{\mathrm{AA}}\rangle} \frac{\mathrm{dN_{AA}/d}p_{\mathrm{T}}}{\mathrm{d}\sigma_{\mathrm{pp}}/\mathrm{d}p_{\mathrm{T}}}, \tag{1}$$

where $\langle T_{\mathrm{AA}}\rangle$ is the nuclear overlap function evaluated in the Glauber model [8], proportional to the number of binary nucleon-nucleon collisions $\langle N_{\mathrm{coll}}\rangle$. By comparing the yields extracted in heavy-ion and pp collisions, this variable tests whether the heavy-ion collision is a mere superposition of independent nucleon-nucleon collisions. If this would be the case, the $R_{\mathrm{AA}}(p_{\mathrm{T}})$ should be 1. Previous measurements showed a yield-suppression at high momenta for unidentified hadrons [9] as well as D mesons [10] and π^0 [11]. This suppression is generally understood as a manifestation of the energy loss the initial partons suffer when they traverse the QGP.

Another observable, that complements the modification of the momentum spectrum is the per-trigger yield obtained from angular correlations:

$$C(\Delta\eta\Delta\varphi) = \frac{1}{N_{\mathrm{trig}}} \frac{\mathrm{dN_{assoc}}}{\mathrm{d}\Delta\eta\mathrm{d}\Delta\varphi}. \tag{2}$$

This analysis considers hadrons with $p_{\mathrm{T}} > 8$ GeV/c as high-momentum triggers. We take the trigger-hadron and investigate the angular distribution of the associated hadrons in different momentum bins. In this way, we can observe how the associated particle distributions are modified by the presence of the medium. When a high-virtuality process occurs it produces, at leading order, two back-to-back partons. If this process occurs close to the surface of the fireball, one of the partons will have to travel a shorter distance through the hot and dense matter than the other. Therefore the correlation analysis suffers a *"surface-bias"* that allows us to study path-length dependence of the parton energy-loss. The high-momentum particles associated with the parton close to the surface are identified as triggers and particles associated with this parton's energy-loss appear on the near-side ($\Delta\varphi \approx 0$). The other parton has to travel through the fireball experiencing large energy-loss and producing associated particles on the away-side ($\Delta\varphi \approx \pi$). To quantify the modification of the particle-yields in a certain momentum interval associated with the trigger, we construct the *per-trigger yield modification factor (I_{AA})*:

$$I_{\mathrm{AA}}(\Delta\varphi, p_{\mathrm{T}}) = \frac{Yield(\text{Pb–Pb})}{Yield(\text{pp})}. \tag{3}$$

Apart from the yield-modification, one can also look at the changes in the shape of the angular distribution of the associated particles produced along the path of the trigger. Such measurement targets the broadening and softening of the initial parton fragmentation and is done by measuring the widths of the near- and away-side peaks.

In the following, we present the results of the analysis on π^0 angular correlation with charged particles ($h^{\pm}$) in pp and Pb–Pb collisions at $\sqrt{s} = 2.76$ TeV per-nucleon and π^0-jet angular correlations in pp collisions at $\sqrt{s} = 7$ TeV. The correlations of D mesons and charged particles in pp collisions at $\sqrt{s} = 7$ TeV are compared with results from p–Pb collisions at $\sqrt{s_{NN}} = 5.02$ TeV and PYTHIA Monte Carlo simulations [12, 13, 14].

2. Analysis setup

The ALICE detector is a powerful system of sub-detectors, that was designed to investigate the properties of QGP in a broad range of variables and kinematics. A detailed description can be found in Ref. [15]. In the presented analysis, the following sub-detectors were used: for the tracking and identification of the charged particles the ITS [16], TPC [17] and TOF [18] were used covering the central pseudo-rapidity region $|\eta| < 0.9$ and full azimuth $\varphi = 2\pi$. The π^0s were reconstructed using the EMCal [19] electromagnetic calorimeter with the acceptance $|\eta| < 0.7$ and $\Delta\varphi = 100°$. In the π^0-jet analysis, jets were reconstructed using the anti$-k_T$ jet-finding algorithm [20]. For the reconstruction only charged tracks have been used in $|\eta| < 0.9$ and jets with $R = 0.4$ and $|\eta| < 0.5$ were accepted.

2.1. Identification of π^0

The neutral pions were identified using the shower-shape topological analysis in the EMCal. For energies larger than $5 - 6$ GeV, the electromagnetic showers produced in the calorimeter by the two photons from π^0 decay merge into a single large cluster of calorimeter cells (a cell is the smallest unit of the calorimeter, with 6×6 cm^2 area at 4.28 m from the interaction point). By studying the energy profile of the cluster, we can determine whether it was produced by a single photon or if it is a superposition of two. This procedure allows us to measure π^0s in the EMCal up to 50 GeV.

2.2. Identification of D mesons

For the D-meson triggered correlation analysis the D^+, D^0 and D^{*+} have been reconstructed in three decay modes: $D^0 \to K^-\pi^+$, $D^+ \to K^-\pi^+\pi^+$ and $D^*(2010)^+ \to D^0\pi^0$ [21]. A detailed description of the procedure can be found in Ref. [22] and the references therein. The selection procedure exploits the displacement of the decay vertex with respect to the primary vertex. Kaons and pions were identified using TPC and TOF detectors and from these the D-meson invariant mass was reconstructed.

2.3. Setup of the correlation studies

In the correlation analysis, the triggers were associated with the unidentified hadrons and charged jets. For each trigger-associated pair, we calculate $\Delta\varphi = \varphi_{\rm trig} - \varphi_{\rm assoc}$ and $\Delta\eta = \eta_{\rm trig} - \eta_{\rm assoc}$ in the signal event. To correct for the detector $\Delta\varphi$ and $\Delta\eta$ acceptance, we divide the raw $(\Delta\varphi\Delta\eta)$ correlation with the mixed-event correlation. This is a correlation, that is not physical and is constructed by taking a trigger from the signal event and correlating it with an associated particle from another, but topologically similar event. Dividing the signal correlation by the mixed, we obtain the pair-acceptance corrected distribution. This distribution consists of two peaks, which sit on top of a pedestal, that is not correlated with the trigger and has different physics origins. In these analyses, we want to only study the per-trigger yield in the peaks so we have to subtract this pedestal. In pp collisions, the background is assumed to be flat and is determined using *Zero Yield At Minimum (ZYAM)* assumption [23]. In Pb–Pb collisions, the situation becomes more complicated, since the background is expected to be modulated by the initial azimuthal anisotropy called elliptic flow [24]. Therefore, the background is subtracted in the analysis of Pb–Pb data using both a flat and a flow assumption.

3. Results and discussion

In this section, we present our results on the angular correlation analyses using π^0 and D mesons as triggers. First, the results of π^0-hadron and π^0-jet correlation are shown, then, we proceed with D-meson analyses. The plots include both systematic and statistical errors, which are not combined but are plotted separately.

IOP Conf. Series: Journal of Physics: Conf. Series **805** (2017) 012011 doi:10.1088/1742-6596/805/1/012011

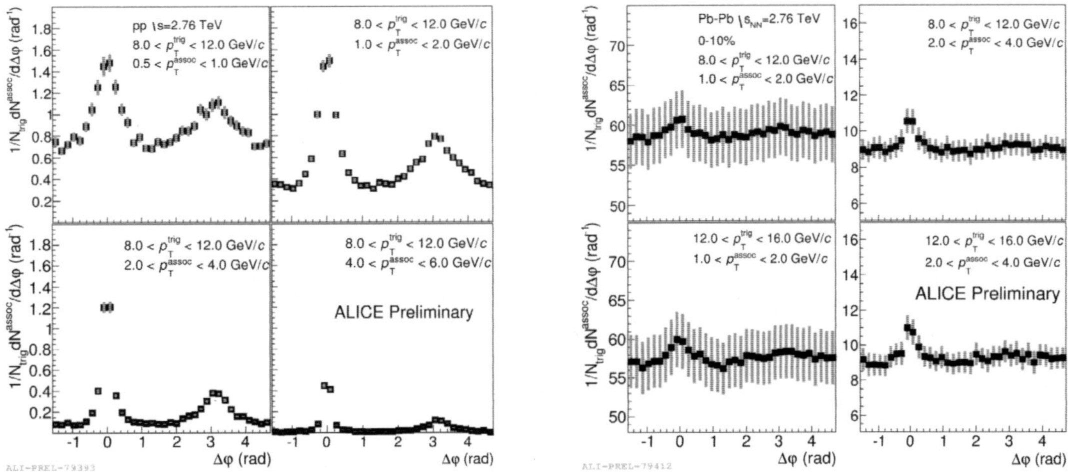

(a) (b)

Figure 1. The π^0 -charged particle correlation: angular distribution of the per-trigger yield for different colliding systems and different momentum of $\pi^0 - h^{\pm}$ pairs without background subtraction. The distributions are corrected for the pair-acceptance. Panel (a) shows results from pp collisions at $\sqrt{s}$ = 2.76 TeV with $8.0 < p_{\mathrm{T}}^{\mathrm{trig}} < 12.0$ GeV/c and $p_{\mathrm{T}}^{\mathrm{assoc}} \in \{(0.5;1.0),(1.0;2.0),(2.0;4.0),(4.0;6.0)\}$ GeV/c. Panel (b) shows the per-trigger yield in Pb–Pb collisions at $\sqrt{s_{\mathrm{NN}}}$ = 2.76 TeV. The momentum ranges of the associated pairs are $p_{\mathrm{T}}^{\mathrm{assoc}} \in \{(1.0;2.0),(2.0;4.0)\}$ GeV/c and the ranges of trigger-momentum are $8.0 < p_{\mathrm{T}}^{\mathrm{trig}} < 12.0$ GeV/c and $12.0 < p_{\mathrm{T}}^{\mathrm{trig}} < 16.0$ GeV/c at top and bottom of the figure, respectively.

3.1. The π^0 triggered correlations

The π^0 - charged particle correlation analysis was performed in pp collisions and 10% most central Pb–Pb collisions at a per-nucleon collision energy of 2.76 TeV.

Figure 1 shows the evolution of the angular distribution of the per-trigger yield as it changes with the momentum of the trigger-associate pair and with colliding system from pp to Pb–Pb collisions. The distributions are corrected for the pair-acceptance but are not corrected for the background below the jet-peaks. We observe that with increasing momentum of the associated particles, this background becomes suppressed and the jet peaks start to become the dominant features in the distribution. This behavior is common to both pp and Pb–Pb systems, but is more pronounced in the former.

Once background is subtracted and only the jet-peaks remain, the per-trigger yield is extracted in the $p_{\mathrm{T}}^{\mathrm{assoc}}$ bins for the near- and away-side. The size of the integration range in $\Delta\varphi$ is 0.7 and is the same on both sides. In the case of pp collisions, only the flat background assumption was considered, whereas in the case of Pb–Pb collisions, a flow background assumption was added. The results are shown in Figure 2. The yield decreases with momentum of the associated tracks and is smaller on the away-side. In the Pb–Pb case, the extracted yields are consistent for both background assumptions within errors.

To quantify the difference between the behavior of the $\pi^0 - h^{\pm}$ correlation in pp and Pb–Pb collisions, the I_{AA} (defined by eq. 3) is calculated from the extracted yields. The $p_{\mathrm{T}}^{\mathrm{assoc}}$ dependence of this variable is shown in Figure 3. On the near-side, we observe a $\approx$ 20% enhancement while on the away-side a $\approx$ 50% suppression. The I_{AA} is independent of momentum and background assumption within errors. The results are compared to the results of $h^{\pm} - h^{\pm}$

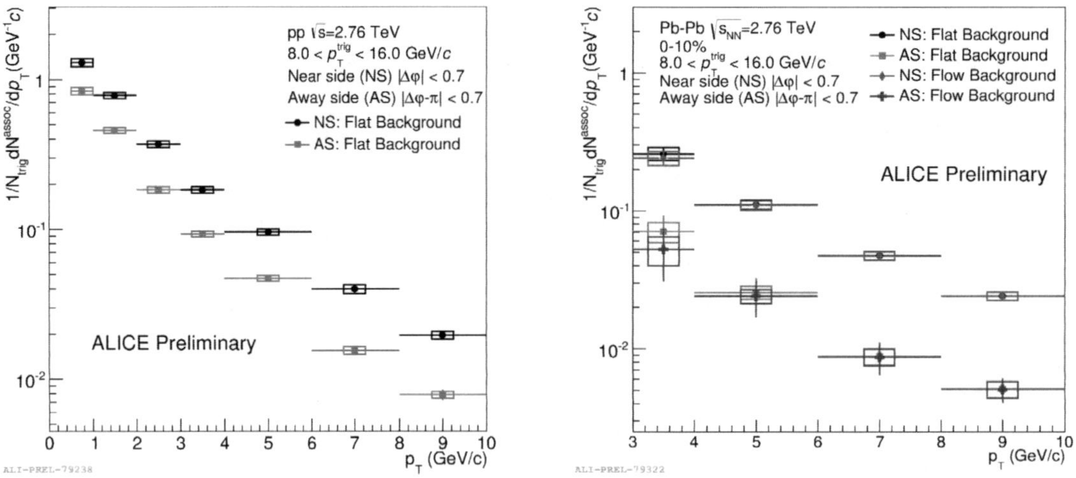

Figure 2. The $\pi^0 - h^{\pm}$ correlation: per-trigger yield on the near- and away-side as a function of associated track momentum in pp collisions at $\sqrt{s} = 2.76$ TeV (left) and in 10% most central Pb–Pb collisions at $\sqrt{s_{\rm NN}} = 2.76$ TeV (right). The background subtracted in pp collisions is flat. In Pb–Pb collisions, the yields are compared for flat and flow background assumption.

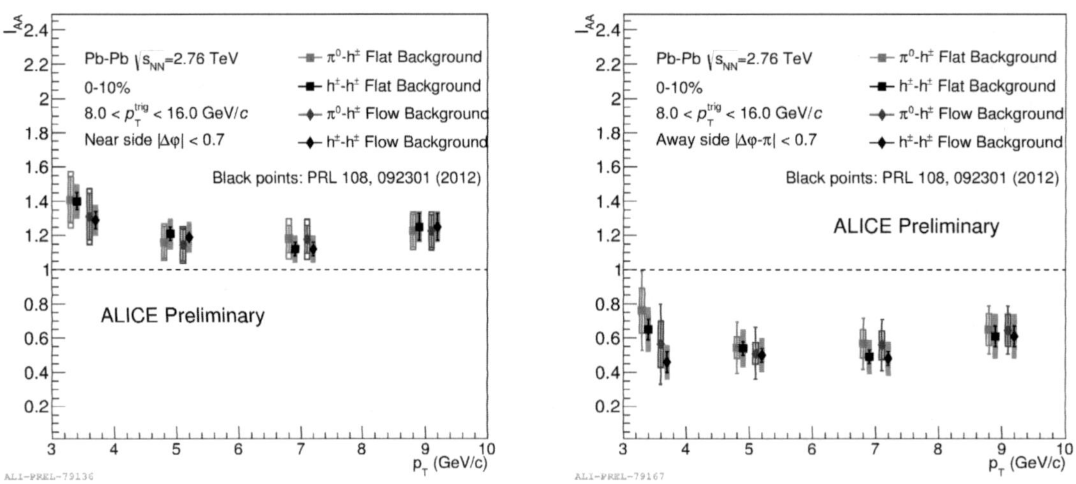

Figure 3. Per-trigger yield modification factor ($I_{\rm AA}$) as a function of the $p_{\rm T}$ of the associated particle on the near-side (left) and away-side (right) for 10% most central Pb–Pb collisions at $\sqrt{s_{\rm NN}} = 2.76$ TeV. The results of the $\pi^0 - h^{\pm}$ analysis are compared to the results of $h^{\pm} - h^{\pm}$ analysis for a flat and a flow background assumption.

analysis [25] and are found to be consistent within errors.

The suppression on the away-side can be understood in terms of medium-induced parton energy-loss mentioned in the introduction. It is interesting to point out, that this suppression is smaller than the one measured by the STAR collaboration at RHIC [26] ($\approx 75\%$). The difference in the suppression might indicate a difference in the biases of the correlation analysis between the two experiments. A discussion on how various biases of the experimental correlation

IOP Conf. Series: Journal of Physics: Conf. Series **805** (2017) 012011 doi:10.1088/1742-6596/805/1/012011

measurements influence the results and their interpretation can be found in [27]. Based on this, the main contributors to the differences may be the interplay between the following effects: *(i)* different "surface bias" of the measurement *(ii)* larger momentum transfer in hard-scattering processes at LHC and *(iii)* change of the fraction of quarks and gluons produced in the final state.

Due to larger collision energies, the partons created in the hard-scatterings have a higher initial energy and it is likely for them to survive the fireball with shifted momentum even if they have to travel through the whole length of the medium. Apart from this, even if the hard-scattering occurs further away from the surface, the partons might be energetic enough to have the particles produced in their fragmentation triggered in the experimental setup. Additionally, the fraction of quarks and gluons created in the final state of hard-scatterings changed when going from RHIC to LHC energies. While the quarks dominated in the RHIC regime, it is more likely to produce gluons at the LHC. Gluons have a softer fragmentation, thus one can expect them to be more prone to interact with the surrounding medium and as a result produce a large number of particles. The higher initial parton energy combined with a softer fragmentation and a higher probability to interact with medium may result in an apparently smaller suppression in a given momentum range.

Looking at the near-side, the STAR collaboration measured an $I_{\rm AA}$ consistent with unity. The consistency with unity suggests, that very little interaction between the trigger and the surrounding matter occurred. The presented ALICE measurement showed a clear enhancement, indicating that the trigger has been affected by its travel through the medium. Following a similar line of thought to the previous paragraph, one can understand this via the difference between the kinematics of the hard-scattering processes between the RHIC and the LHC. Apart from identifying triggers at the surface, one identifies also the ones originating inside the fireball. These suffer the medium-induced energy-loss and are identified as triggers in a lower momentum bin together with an abundance of particles around them, that were produced as the result of the parton-medium interaction. The fact supporting that the difference in the $I_{\rm AA}$ between STAR and ALICE has the same source on the near- and away-side is that ALICE adds $\approx 20\%$ to both sides of the STAR result.

Other possible interpretation of this enhancement is the medium-induced softening of the fragmentation function, discussed briefly in Ref. [25]. Hadrons in a given momentum range, which were produced in such modified fragmentation, originate from higher-momentum partons in Pb–Pb collisions as compared to pp. This difference between the two systems is then translated into $I_{\rm AA} > 1$.

To further test these ideas, one would need to conduct an analysis targeting directly the jet-fragmentation via the measurements of hadron-jet correlations. Such measurements have already started. The π^0-jet correlation analysis was performed in pp collisions at $\sqrt{s} = 7$ TeV. Three trigger-momentum bins and three jet-momentum thresholds were studied. The angular distributions of the per-trigger jet-yield are shown in Figure 4. We observe clear peaks on both the near- and away-side. The jet-yield increases with the momentum of π^0 and decrease with the jet-momentum threshold.

Figure 5 shows the widths of the peaks and their dependence on $p_{\rm T}^{\rm trig}$ and $p_{\rm T,ch.jet}^{\rm assoc}$. Within errors, the widths are independent of associated jet momentum-threshold and they exhibit a decreasing trend with the trigger-particle momentum.

These results show that the meson is produced closer to the jet-axis as its momentum increases. On the other hand, the width of the angular distribution seems to be independent of jet's threshold momentum. This indicates a small jet-momentum dependence of jet-fragmentation into π^0s. However, it is important to mention that this measurement was done using only charged jets, and so it is difficult to establish a direct relationship between the charged-jet and neutral-pion momentum. A more robust claim on the jet's fragmentation into

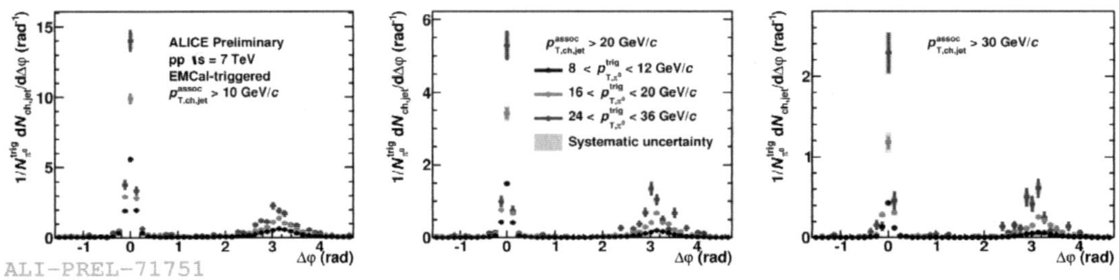

Figure 4. Angular distribution of per-trigger jet-yield from the π^0-jet correlation analysis in pp collisions at $\sqrt{s} = 7$ TeV. The distributions are compared for three trigger-momentum bins: $p^{\text{trig}}_{\text{T,ch.jet}} \in \{(8;12), (16;20), (24;36)\}$ GeV/c. For the associated jet three momentum thresholds are defined: $p^{\text{assoc}}_{\text{T,ch.jet}} > 10$ GeV/c (left), $p^{\text{assoc}}_{\text{T,ch.jet}} > 20$ GeV/c (middle) and $p^{\text{assoc}}_{\text{T,ch.jet}} > 30$ GeV/c (right).

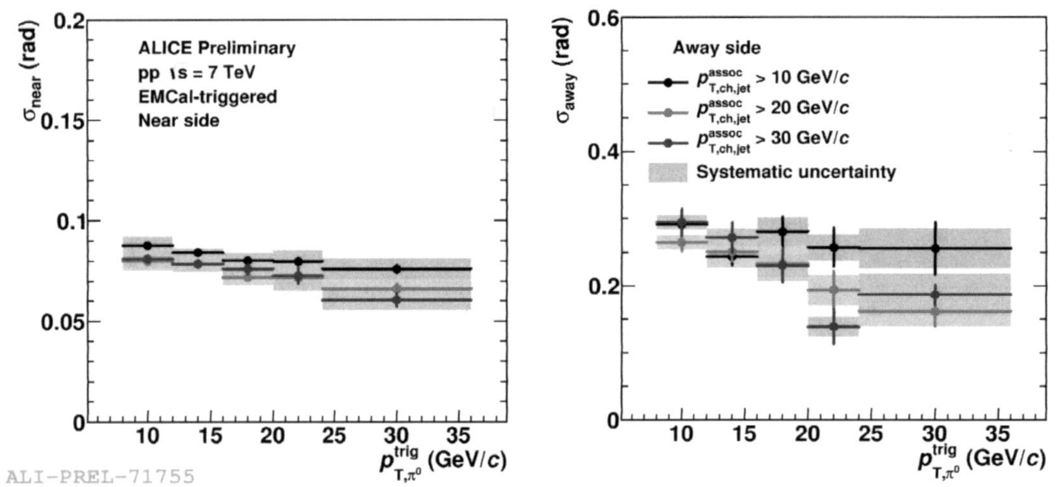

Figure 5. Widths of the angular distribution of per-trigger jet-yield from the π^0-jet analysis in pp collisions at $\sqrt{s} = 7$ TeV. The widths are plotted as a function of π^0 momentum for three associated jet-momentum thresholds: $p^{\text{assoc}}_{\text{T,ch.jet}} > 10, 20, 30$ GeV/c. The widths are extracted on the near- (left) and away-side (right).

π^0s requires a full-jet reconstruction with both the charged and the neutral part.

3.2. D meson - charged particle azimuthal correlations

Azimuthal correlations of D mesons and charged particles were studied in pp collisions at $\sqrt{s} = 7$ TeV and p–Pb collisions at $\sqrt{s_{NN}} = 5.02$ TeV. The D mesons were identified in the range $8 < p^D_T < 16$ GeV/c and were correlated to unidentified charged particles with $p^{\text{assoc}}_T > 1$ GeV/c. The per-trigger yield as a function of $\Delta\varphi$ is shown in Figure 6. The distributions are compared to PYTHIA Monte Carlo simulations of pp collisions at corresponding center-of-mass energies: PYTHIA 8 [13] and PYTHIA 6 with Perugia 2010 and Perugia 2011 tunes [14]. Both pp and p–Pb results are within errors reproduced by the Monte Carlo models.

Figure 7 shows a direct comparison of angular distribution of per-trigger yield (left) and

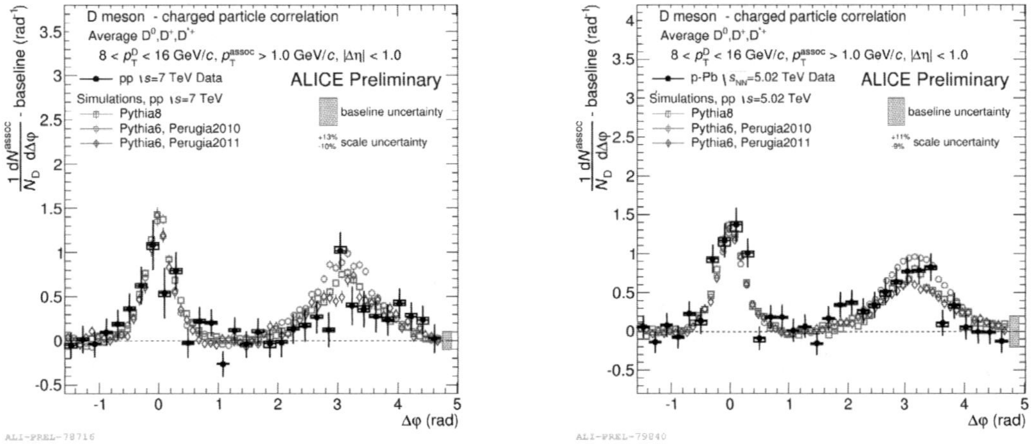

Figure 6. The D meson - charged particle azimuthal correlations:: angular distribution of per-trigger yield in pp collisions at $\sqrt{s} = 7$ TeV (left) and p–Pb collisions at $\sqrt{s_{\rm NN}} = 5.02$ TeV (right). The trigger used is a D meson with transverse momentum $8.0 < p_{\rm T}^{\rm D} < 16$ GeV/c. The associated tracks are unidentified particles with $p_{\rm T}^{\rm assoc} > 1.0$ GeV/c. The distributions are compared to Monte Carlo simulations of pp collisions at $\sqrt{s} = 7$ TeV in the case of pp collisions, and $\sqrt{s} = 5.02$ TeV in the case of p–Pb collisions. The PYTHIA 8 and PYTHIA 6 generators are used, the latter implementing two tunes: Perugia 2010 and Perugia2011.

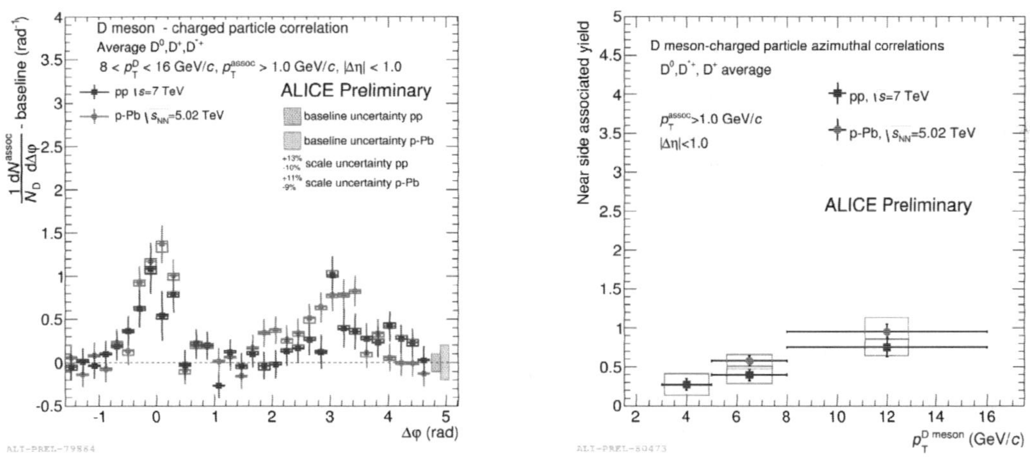

Figure 7. Comparison of D meson - charged particle correlations in pp collisions at $\sqrt{s} = 7$ TeV and p–Pb collisions at $\sqrt{s} = 5.02$ TeV. Left: Comparison of the angular distribution of per-trigger yield. Right: Comparison of near-side per-trigger yield as a function D-meson momentum.

near-side yield as a function of $p_{\rm T}^{D}$ (right) in pp and p–Pb collisions. The results obtained in the two collision systems agree within errors. However, due to large uncertainties it is difficult to draw strong conclusions about the comparison.

IOP Conf. Series: Journal of Physics: Conf. Series **805** (2017) 012011 doi:10.1088/1742-6596/805/1/012011

4. Conclusions and final remarks

We have presented an overview of the results on the correlation-measurements with π^0 and D mesons as high-p_T triggers. We can conclude, that the results from π^0-triggered correlations in Pb–Pb collisions are consistent with the presence of hot and dense matter that causes partons to lose energy as they travel through it. This interaction is affected by the kinematics of the partons created in the initial stages of the collision as can be seen from the comparison of RHIC and LHC results. A ore detailed investigation of the jet-fragmentation modification has already begun.

In the D-meson correlation measurements, first results are available in pp and p–Pb collisions, setting an important baseline for upcoming heavy-ion analyses. In p–Pb collisions, no indication of hot nuclear effects were observed. In the upcoming Run II, the statistical uncertainties are expected to significantly improve. The ITS upgrade planned for Run III should provide an improved spatial resolution for the secondary vertex recognition, which is crucial for the performance in heavy-flavor measurements.

4.1. Acknowledgments

I would like to thank the ALICE Collaboration for giving me the opportunity to present the results. Further, I would like to thank the Wigner RCP and the OTKA grant NK 116109 for providing necessary funding for my attendance of the 10th International Workshop on High-pT Physics in the RHIC/LHC era in Nantes, France (9th – 12th September 2014).

References

[1] Bazavov A *et al.* *Phys.Rev.*, D **85**:054503, 2012.
[2] Alessandro B *et al.* *J.Phys.*, G **32**:1295–2040, 2006.
[3] Gyulassy M and Plumer M. *Phys.Lett.*, B **243**:432–438, 1990.
[4] Baier R *et al.* *Nucl.Phys.*, B **484**:265–282, 1997.
[5] Thoma H M and Gyulassy M. *Nucl.Phys.*, B **351**:491–506, 1991.
[6] Braaten E and Thoma M H. *Phys.Rev.*, D **44**:1298–1310, 1991.
[7] Dokshitzer Y L and Kharzeev D E. *Phys.Lett.*, B **519**:199–206, 2001.
[8] Miller M L *et al.* *Ann.Rev.Nucl.Part.Sci.*, **57**:205–243, 2007.
[9] Aamodt K *et al.* *Phys.Lett.*, B **696**:30–39, 2011.
[10] Abelev B *et al.* *JHEP*, **1209**:112, 2012.
[11] Abelev B *et al.* *Eur.Phys.J.*, C **74**(10):3108, 2014.
[12] Mrenna S Sjostrand T and Skands P Z. *JHEP*, **0605**:026, 2006.
[13] Sjstrand T *et al.* 2014.
[14] Skands P Z. *Phys.Rev.*, D **82**:074018, 2010.
[15] Aamodt K *et al.* *JINST*, **3**:S08002, 2008.
[16] Aamodt K *et al.* *JINST*, **5**:P03003, 2010.
[17] Alme J *et al.* *Nucl.Instrum.Meth.*, A **622**:316–367, 2010.
[18] Akindinov A *et al.* *Nucl.Instrum.Meth.*, A **615**:37–41, 2010.
[19] Bourrion O *et al.* *JINST*, **8**:C01013, 2013.
[20] and Salam G P Cacciari M and Soyez G. *Eur.Phys.J.*, C **72**:1896, 2012.
[21] Nakamura K *et al.* *J.Phys.*, G **37**:075021, 2010.
[22] Abelev B *et al.* *JHEP*, **1201**:128, 2012.
[23] Ajitanand N N *et al.* *Phys.Rev.*, C **72**:011902, 2005.
[24] Adler C *et al.* *Phys.Rev.Lett.*, **90**:032301, 2003.
[25] Aamodt K *et al.* *Phys.Rev.Lett.*, **108**:092301, 2012.
[26] Adams J *et al.* *Phys.Rev.Lett.*, **97**:162301, 2006.
[27] Renk T. *Phys.Rev.*, C **88**(5):054902, 2013.

Detailed study of parton energy loss via measurement of fractional momentum loss of high p_{T} hadrons in heavy ion collisions

Takao Sakaguchi, for the PHENIX Collaboration

Brookhaven National Laboratory, Physics Department, Upton, NY 11973-5000, USA

E-mail: takao@bnl.gov

Abstract. PHENIX measurement of the fractional momentum loss ($\delta p_{\mathrm{T}}/p_{\mathrm{T}}$) of high p_{T} identified hadrons are presented. The $\delta p_{\mathrm{T}}/p_{\mathrm{T}}$ of high p_{T} π^0 which are computed from 39 GeV Au+Au up to 2.76 TeV Pb+Pb are found to vary by a factor of six. We plotted the $\delta p_{\mathrm{T}}/p_{\mathrm{T}}$ against several global variables, N_{part}, N_{qp} and $dN_{\mathrm{ch}}/d\eta$, and found global features. It was found that 200 GeV Au+Au points are merging into the central 2.76 TeV Pb+Pb points when plotting $\delta p_{\mathrm{T}}/p_{\mathrm{T}}$ against $dN_{\mathrm{ch}}/d\eta$.

1. Introduction

The interaction of hard scattered partons with the medium created by heavy ion collisions (i.e., quark-gluon plasma, QGP) has been of interest since the beginning of the RHIC running. A large suppression of the yields of high transverse momentum (p_{T}) hadrons which are the fragments of such partons was observed, suggesting that the matter is sufficiently dense to cause parton-energy loss prior to hadronization [1]. The PHENIX experiment [2] has been exploring the highest p_{T} region with single π^0 mesons which are leading hadrons of jets. We show a calculation of the energy loss of partons published almost 20 years ago in Fig. 1(a) [3]. Although the measurement of the momentum shift is the ultimate goal, the paper suggested looking at the ratio of the high p_{T} single hadrons in Au+Au and p+p collisions as an alternate way. Since then, most of the experiments including PHENIX have looked at the nuclear modification factors, R_{AA} ($\equiv (dN_{\mathrm{AA}}/dydp_{\mathrm{T}})/(\langle T_{\mathrm{AA}}\rangle d\sigma_{pp}/dydp_{\mathrm{T}})$), and quantified the energy loss effect via its suppression. In the present analysis, we evaluate the momentum shift of high p_{T} hadrons instead of R_{AA}.

2. Fractional momentum loss $\delta p_{\mathrm{T}}/p_{\mathrm{T}}$

With a larger statistics of both $p+p$ and Au+Au data recently collected, it became possible to measure the momentum shift directly. Fig. 1(b) depicts the method to compute such shift. We have statistically extracted the fractional momentum loss ($\delta p_{\mathrm{T}}/p_{\mathrm{T}}$, $\delta p_{\mathrm{T}} \equiv p_{\mathrm{T}} - p_{\mathrm{T}}'$, where p_{T} is the transverse momentum of the $p+p$ data, and p_{T}' is that of the Au+Au data) of the of the hadrons using their p_{T} spectra measured in $p+p$ and Au+Au collisions [4]. Since the number of data points is finite, a fit to scaled $p+p$ is needed to evaluate $\delta p_{\mathrm{T}}/p_{\mathrm{T}}$ at a given Au+Au invariant yield. The uncertainty of the $\delta p_{\mathrm{T}}/p_{\mathrm{T}}$ is calculated by converting the quadratic sum of the uncertainties on the yields of Au+Au and $p+p$ points, using the $p+p$ fit function. Statistical

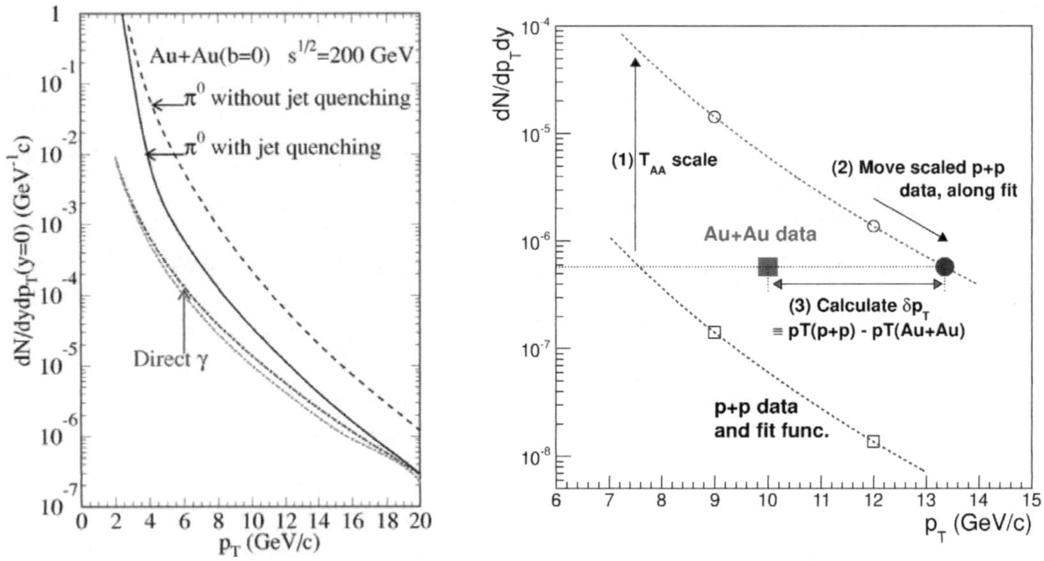

Figure 1. (a, left) A calculation demonstrating that the suppression of the π^0 yield can be modeled by a relative transverse momentum shift of the transverse momentum spectra in A+A collisions. (b, right) Method of calculating average $\delta p_T/p_T$. We scale the differential cross-section measured in $p+p$ by T_{AA} (yields are scaled by N_{coll} corresponding to the centrality class of Au+Au data, shift the $p+p$ points closest to Au+Au in yield, and calculate the momentum difference of $p+p$ and Au+Au points.

and systematic uncertainties are individually calculated in the same way. The uncertainties on T_{AA} and p+p luminosity are not plotted but indicated in the legend. Fig. 2(a) show the R_{AA} for the π^0's in Au+Au collisions at $\sqrt{s_{NN}}$ =200 GeV from the RHIC Year-7 run. Using this method, we obtained the $\delta p_T/p_T$ for the same dataset as shown in Fig. 2(b). Similarly, the R_{AA} for the π^0's in 0-10 % Au+Au collisions at $\sqrt{s_{NN}}$ =39, 62 and 200 GeV from the RHIC Year-7 and Year-10 runs shown in Fig. 3(a) are replotted in the form of $\delta p_T/p_T$ as shown in Fig. 3(b). The R_{AA}'s don't change as much as the change in cms energies as seen in Fig. 3(a). However, when we look at $\delta p_T/p_T$ for the corresponding dataset, we found that the $\delta p_T/p_T$ changes by a factor of three from 39 to 200 GeV as shown in Fig. 3(b) [5].

On the other hand, the R_{AA}'s look similar between RHIC and LHC (Fig. 4(a)). The $\delta p_T/p_T$ is found to change by a factor of $\sim$ 1.5 from 200 to 2.76 TeV (Fig. 4(b)). To summarize, even the R_{AA}'s don't change as much as the change in the cms energies, the $\delta p_T/p_T$'s show a factor of six variation from 39 GeV to 2.76 TeV. This fact has not been found by looking at R_{AA}.

3. Scaling property of $\delta p_T/p_T$

In order to study the systematics of $\delta p_T/p_T$, we plot the $\delta p_T/p_T$ against several global variables such as N_{part}, N_{qp} (number of quark participants) and $dN_{ch}/d\eta$. We first plotted the $\delta p_T/p_T$ against N_{part} as shown in Fig. 5. All the plots shown in this section are at $p_T(p+p) = 7$ GeV/c in order to reach the hard scattering regime. For the N_{part} dependence we see that the Cu+Cu and Au+Au are nicely lined up, implying that for the same cms energy, the $\delta p_T/p_T$ scales with N_{part}. This is consistent with the fact that R_{AA} is similar at the same N_{part} between Cu+Cu and

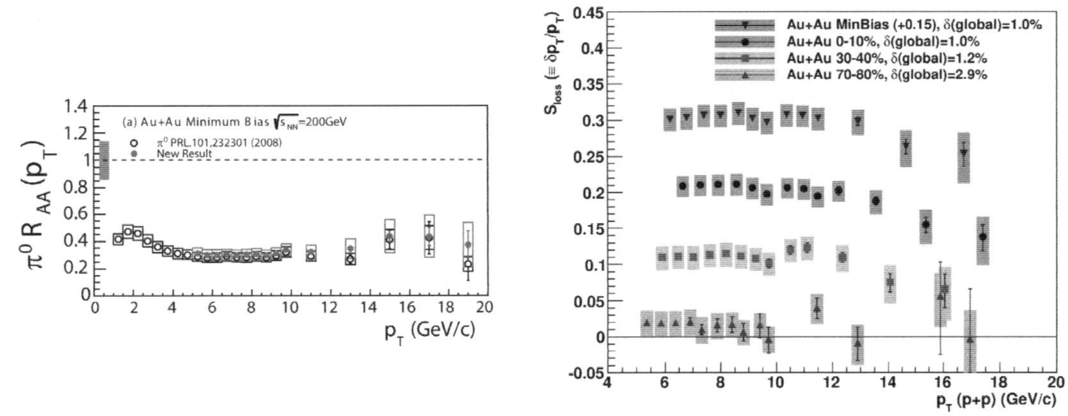

Figure 2. (a, left) R_{AA} of π^0's for 200 GeV Au+Au collisions obtained from RHIC Year-7 run. The gray bars around the unity shows the quadratic sum of p+p luminosity and T_{AA} uncertainties. (b, right) $\delta p_{\mathrm{T}}/p_{\mathrm{T}}$ for the same dataset.

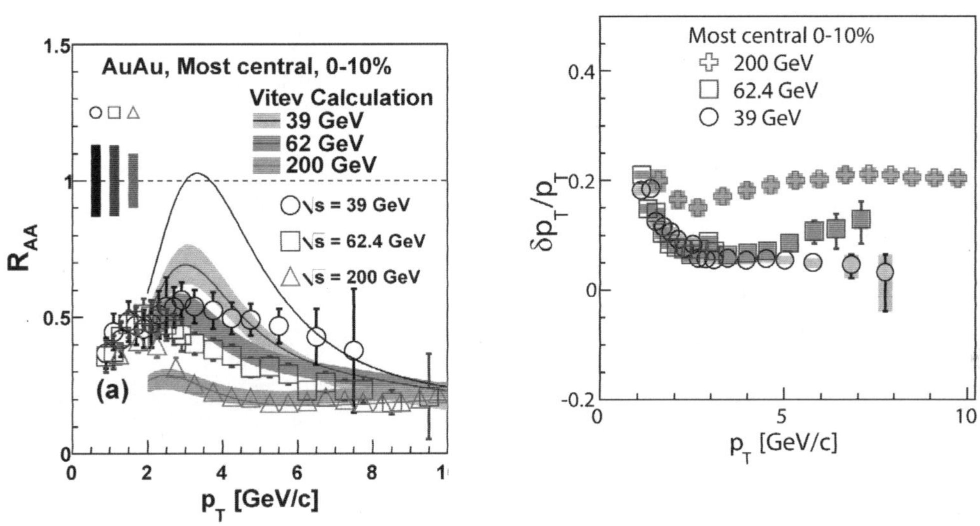

Figure 3. (a, left) R_{AA} of π^0's for 39, 62 in 0-10 % Au+Au collisions obtained from RHIC Year-10 run and in 0-10 % 200 GeV Au+Au collisions from RHIC Year-10 run. The bands around the unity shows the quadratic sum of p+p luminosity and T_{AA} uncertainties for each cms energies. (b, right) $\delta p_{\mathrm{T}}/p_{\mathrm{T}}$ for the same dataset.

Au+Au collisions [7]. The Pb+Pb points are consistently off the trend of 200 GeV points, but the slopes of both systems look similar. Fig. 6 shows $\delta p_{\mathrm{T}}/p_{\mathrm{T}}$ against N_{qp}. The detail description of how the number of quark participants are obtained can be found in the literature [8]. We employed a Monte-Carlo-Glauber (MC-Glauber) model to calculate the numbers. We first determine the quark-quark inelastic cross section ($\sigma_{qq}^{\mathrm{inel}}$) for each collision energy such that the inelastic nucleon-nucleon cross section ($\sigma_{NN}^{\mathrm{inel}}$) is reproduced. Then the model is modified to

IOP Conf. Series: Journal of Physics: Conf. Series **805** (2017) 012012 doi:10.1088/1742-6596/805/1/012012

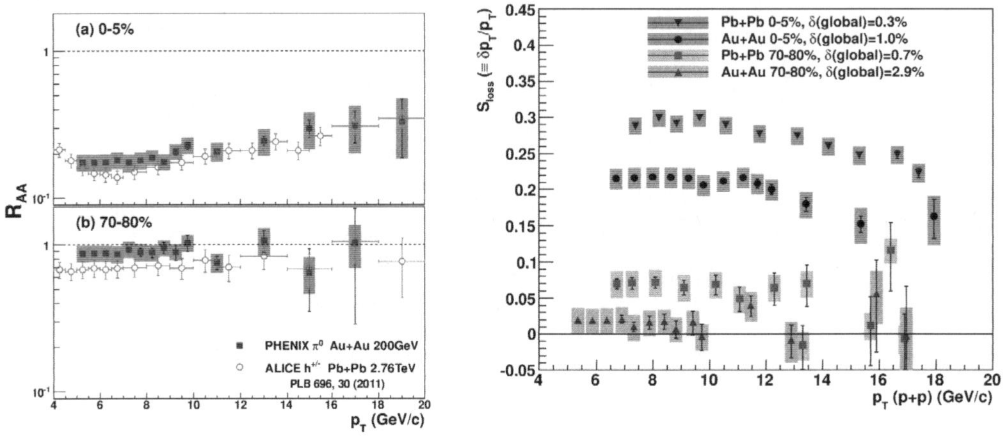

Figure 4. (a, left) R_{AA} of π^0's for 200 GeV Au+Au collisions obtained from RHIC Year-7 run and charged hadrons for 2.76 TeV Pb+Pb collisions obtained by the ALICE experiment at LHC [6]. (b, right) $\delta p_T/p_T$ for the same dataset.

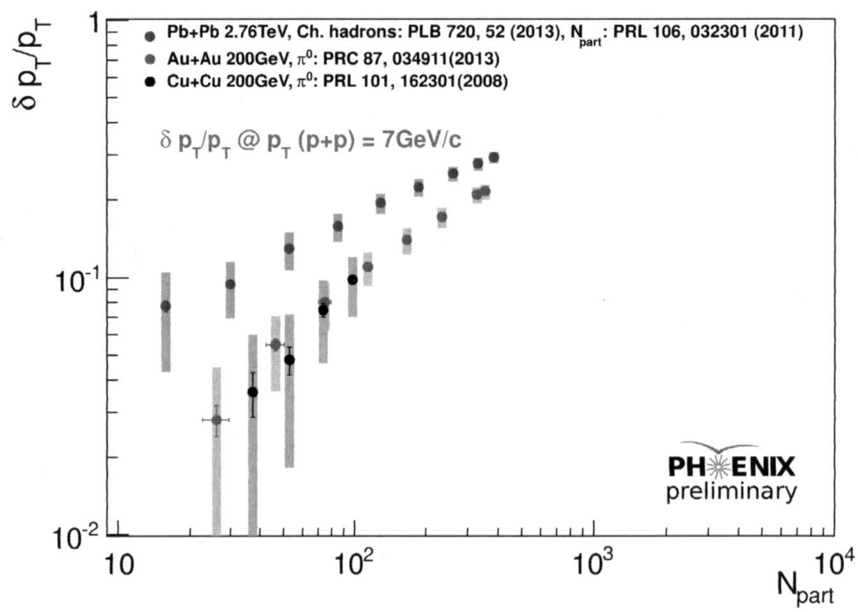

Figure 5. $\delta p_T/p_T$ as a function of N_{part} for π^0's in 200 GeV Au+Au and Cu+Cu collisions measured by PHENIX and charged hadrons in 2.76 TeV Pb+Pb collisions measured by ALICE.

handle the quark-quark rather than nucleon-nucleon collisions. The nuclei are placed according to a Woods-Saxon distribution and then three quarks are distributed around the center of each nucleon following the distribution of:

$$\rho^{proton}(r) = \rho_0^{proton} \times e^{-ar}$$

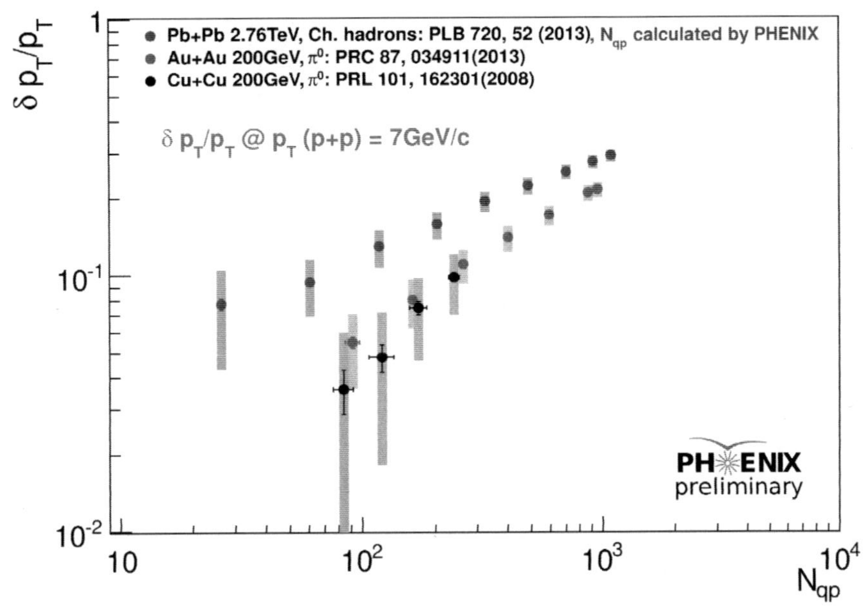

Figure 6. $\delta p_T/p_T$ as a function of N_{qp} for π^0's in 200 GeV Au+Au and Cu+Cu collisions measured by PHENIX and charged hadrons in 2.76 TeV Pb+Pb collisions measured by ALICE.

where $a = \sqrt{12}/r_m = 4.27\,\text{fm}^{-1}$ and $r_m = 0.81$ fm is the rms charge radius of the proton. A pair of quarks, one from each nucleus, interact with each other if their distance d in the plane transverse to the beam axis satisfies the condition of $d < \sqrt{\sigma_{qq}^{\text{inel}}/\pi}$. The number of quark participants as a function of the number of nucleon participants is nonlinear, especially for low values of N_{part}. In Fig. 6, the similar feature as the previous plot is seen. Since the N_{part} is a factor of 2-3 higher than N_{qp}, all the points are systematically moved to the right. We expected that the N_{qp} dependence show a different trend compared to N_{part} dependence since the N_{qp} involves the partonic degree of freedom, which should play a role in the energy loss mechanism. However, it turned out that trend is very similar between N_{part} and N_{qp} dependence. Finally, we plotted the $\delta p_T/p_T$ against the charged multiplicity, $dN_{\text{ch}}/d\eta$, as shown in Fig. 7. In this plot, we added one 62 GeV Au+Au point which is 0-10 % centrality. We expect that $dN_{\text{ch}}/d\eta$ well represents the energy density of the system. It is interesting to note that the most central Au+Au 200 GeV points tend to merge into the most central points of Pb+Pb collisions, while they deviate from each other as going to lower $dN_{\text{ch}}/d\eta$. This systematic trend has not been found by looking at R_{AA}'s which look similar across the systems. In order to cross-check this new result, we have performed a power-law fit to $\delta p_T/p_T$ vs $dN_{\text{ch}}/d\eta$ points from 200 GeV Au+Au collisions, and compared the power with the result obtained from a different method [9]. We fitted the points of this work with $\delta p_T/p_T = \beta(dN_{\text{ch}}/d\eta)^{\alpha/1.19}$ assuming $dN_{\text{ch}}/d\eta \propto (N_{\text{part}})^{1.19}$ [10], and obtained α as 0.64±0.07. Assuming the spectra shape follows a power-law with the power n, one can write the relation between $\delta p_T/p_T$ and R_{AA} as:

$$S_{\text{loss}} \equiv \delta p_T/p_T = \beta N_{\text{part}}{}^{\alpha}, R_{\text{AA}} = (1 - S_{\text{loss}})^{n-2} = (1 - \beta N_{\text{part}}{}^{\alpha})^{n-2}$$

Following this relation, we obtained the power α as 0.57±0.13 from the fit to the integrated R_{AA} as a function of N_{part} in the literature [9]. We therefore confirmed that the powers obtained by the two methods are consistent.

IOP Conf. Series: Journal of Physics: Conf. Series **805** (2017) 012012 doi:10.1088/1742-6596/805/1/012012

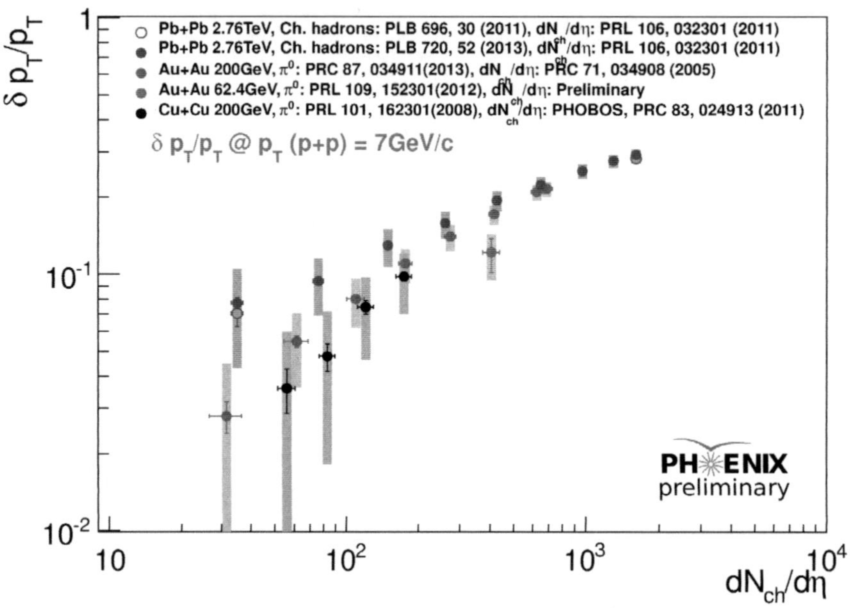

Figure 7. $\delta p_T/p_T$ as a function of $dN_{ch}/d\eta$ for π^0's in Au+Au collisions at 200 GeV and 62.4 GeV, and in Cu+Cu collisions at 200 GeV measured by PHENIX and charged hadrons in 2.76 TeV Pb+Pb collisions measured by ALICE.

4. Summary

We presented PHENIX measurement of the fractional momentum loss ($\delta p_T/p_T$) of high p_T identified hadrons. By looking at the $\delta p_T/p_T$ instead of R_{AA}, we found many interesting features. The $\delta p_T/p_T$ of high p_T π^0 which are computed from 39 GeV Au+Au over to 2.76 TeV Pb+Pb are found to vary by a factor of six. We plotted the $\delta p_T/p_T$ against several global variables, N_{part}, N_{qp} and $dN_{ch}/d\eta$. It was found that 200 GeV Au+Au points are merging into the central 2.76 TeV Pb+Pb points when plotting $\delta p_T/p_T$ against $dN_{ch}/d\eta$. We performed a power-law fit to the $\delta p_T/p_T$ vs $dN_{ch}/d\eta$, and obtained a power that is consistent with the one obtained from the fit to the integrated R_{AA}. We are going to add points from other systems to systematically investigate the $\delta p_T/p_T$.

References

[1] S. S. Adler *et al.* [PHENIX Collaboration], Phys. Rev. Lett. **91**, 072301 (2003).
[2] K. Adcox *et al.* [PHENIX Collaboration], Nucl. Instrum. Meth. A **499**, 469 (2003).
[3] X. -N. Wang, Phys. Rev. C **58**, 2321 (1998).
[4] A. Adare *et al.* [PHENIX Collaboration], Phys. Rev. C **87**, 034911 (2013).
[5] A. Adare *et al.* [PHENIX Collaboration], Phys. Rev. Lett. **109**, 152301 (2012).
[6] K. Aamodt *et al.* [ALICE Collaboration], Phys. Lett. B **696**, 30 (2011).
[7] A. Adare *et al.* [PHENIX Collaboration], Phys. Rev. Lett. **101**, 162301 (2008) [arXiv:0801.4555 [nucl-ex]].
[8] S. S. Adler *et al.* [PHENIX Collaboration], Phys. Rev. C **89**, no. 4, 044905 (2014).
[9] A. Adare *et al.* [PHENIX Collaboration], Phys. Rev. Lett. **101**, 232301 (2008).
[10] S. S. Adler *et al.* [PHENIX Collaboration], Phys. Rev. C **71**, 034908 (2005) [Erratum-ibid. C **71**, 049901 (2005)].

IOP Conf. Series: Journal of Physics: Conf. Series **805** (2017) 012013 doi:10.1088/1742-6596/805/1/012013

Study of high-p_{T} hadron-jet correlations in ALICE

Filip Krizek for the ALICE Collaboration
Nuclear Physics Institute AS CR, Hlavní 130, 25068 Řež u Prahy, Czech Republic
E-mail: filip.krizek@cern.ch

Abstract. We report the measurement of semi-inclusive p_{T} spectra of charged particle jets that recoil from a high-p_{T} hadron trigger in Pb–Pb and pp collisions at $\sqrt{s_{\mathrm{NN}}} = 2.76$ TeV and $\sqrt{s} = 7$ TeV, respectively. In this analysis, the copious yield of uncorrelated trigger hadron-jet matchings in central Pb–Pb collisions is removed by calculating the difference between two spectra corresponding to exclusive trigger hadron p_{T} ranges. This procedure does not impose any fragmentation bias on the recoil jet population, which is therefore collinear and infrared safe. The resulting distributions obtained for different values of jet resolution parameter are used to study the modification of jet structure in the medium.

1. Introduction

Collisions of ultrarelativistic heavy nuclei produce an exotic state of matter which is believed to consist of deconfined quarks and gluons. Strong pressure gradients present at the initial collision stage lead to a rapid collective expansion of the created medium during which quarks and gluons form colorless hadrons. Jets and high-p_{T} hadrons have many advantageous properties to be considered as well suited probes of this matter. First of all they originate from hard parton-parton interactions that occur during early stages of the collision. These scattered partons are known to undergo large energy losses when penetrating through the medium which has direct impact on the energies of outgoing jets and high-p_{T} hadrons. As a result of this parton-medium interaction we observe the so called jet quenching phenomenon manifested by highly p_{T} imbalanced pair of jets.

Heavy-ion collisions result usually in events where the multiplicity of final-state particles is high. Thus, hard scattering products are often embedded into a densely populated, soft, and fluctuating background. Under these conditions, jet finding algorithms frequently create also combinatorial jets composed of background particles only. In principle, the number of these combinatorial jets can be reduced by imposing an additional requirement to have a high-p_{T} leading track within the reconstructed jet. Such constraint, however, breaks down the infrared safety and biases the jet fragmentation. Consequently, this can have an obvious unwanted impact especially on the quenched jets that resemble the combinatorial background jets.

As suggested in [1] hadron-jet coincidence measurements offer a way to remove the contribution of combinatorial jets at the event-ensemble level without introducing a bias on the jet population. Following this idea we report on the analysis of p_{T} spectra of jets associated with a high-p_{T} trigger hadron in Pb–Pb collisions at a center-of-mass energy per nucleon pair $\sqrt{s_{\mathrm{NN}}} = 2.76$ TeV.

IOP Conf. Series: Journal of Physics: Conf. Series **805** (2017) 012013 doi:10.1088/1742-6596/805/1/012013

2. Data analysis

We analyzed 9 M events (10% most central events) and 3 M events (20% most central events) of Pb–Pb collisions at $\sqrt{s_{\mathrm{NN}}} = 2.76$ TeV measured by the ALICE detector [2] in 2010 and 2011, respectively. Accepted charged tracks had pseudorapidity range $|\eta| < 0.9$ and $p_{\mathrm{T}} > 150$ MeV/c. Jet reconstruction uses the charged tracks only and is based on the anti-$\mathrm{k_t}$ algorithm [3]. The imposed infrared cutoff is given by the low p_{T} cut on tracks. Three values of the resolution parameter were employed, $R = 0.2$, 0.4 and 0.5. In order to have all jets fully contained within the ALICE acceptance the centroids of reconstructed jets were constrained to be within the pseudorapidity interval $|\eta_{\mathrm{jet}}| < 0.9 - R$.

The background energy density ρ is estimated with the area based method [4]. The reconstructed jet p_{T} is then event by event corrected for the underlying event activity estimated by the product of the calculated background density and jet area.

The raw p_{T} spectrum of jets has to be corrected for the momentum smearing induced by the fluctuating background and for detector response. The corresponding response matrix that relates the p_{T} of matched reconstructed jets on the particle and detector level is assumed to factorize into a product of matrices that describe both effects. The response matrix was inverted and regularized by means of the SVD [5] and Bayes' theorem [6] based algorithms implemented in the RooUnfold package [7] and was used to correct the raw spectra.

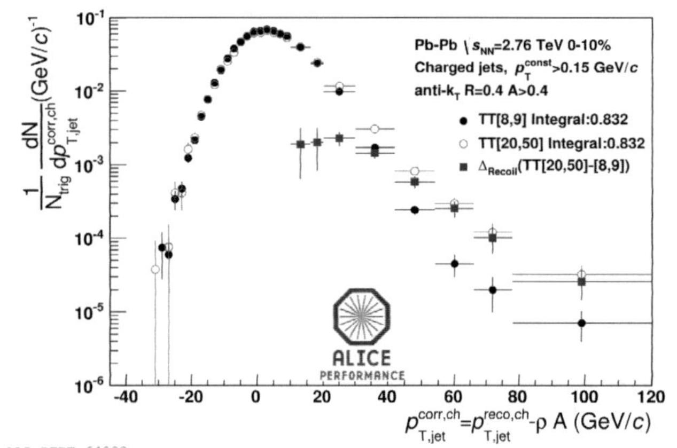

Figure 1. Raw p_{T} spectra of recoil jets associated with two exclusive trigger hadron p_{T} bins, [8,9] GeV/c and [20,50] GeV/c. The corresponding Δ_{recoil} distribution was calculated based on Eq. (1). See text for more details.

The presented analysis is based on jets that are oriented nearly back-to-back in azimuth w.r.t. a high-p_{T} trigger hadron, namely $|\varphi_{\mathrm{trig}} - \varphi_{\mathrm{jet}} - \pi| < 0.6$ rad. The benefit of this choice follows from the fact that the presence of the outgoing high-p_{T} trigger hadron biases the hard scattering to be located close to the surface and the mother parton to be directed toward the outside of the collision zone [8]. Hence, the recoiling jet will on average travel longer distance through the medium. Figure 1 shows background corrected, per trigger normalized raw p_{T} spectra of recoil jets associated with two exclusive trigger hadron p_{T} ranges, [8, 9] GeV/c and [20, 50] GeV/c. The trigger hadrons from the p_{T} bin [20, 50] GeV/c originate on average from hard scattering processes with larger Q^2 than the trigger hadrons from [8, 9] GeV/c. Therefore the associated recoil jets in the first case have also a harder p_{T} spectrum than in the latter case. The positive part of the spectrum is, however, populated also by random matchings of trigger hadron with background jets and the data do not show a clear separation between the contributions of true and accidental coincidences. In the negative part of the spectrum, accidental combinations of trigger hadron with background jets dominate and the jet yield per trigger is to a good approximation independent of the trigger p_{T}. Based on p_{T} spectra of recoil jets associated with two exclusive trigger hadron p_{T} ranges we define

$$\Delta_{\text{recoil}} = \frac{1}{N_{\text{trig}}} \frac{\mathrm{d}N_{\text{jet}}}{\mathrm{d}p_{\text{T,jet}}}\bigg|_{p_{\text{T,trig}} \in [20,50]} - \frac{1}{N_{\text{trig}}} \frac{\mathrm{d}N_{\text{jet}}}{\mathrm{d}p_{\text{T,jet}}}\bigg|_{p_{\text{T,trig}} \in [8,9]} \quad (1)$$

where N_{trig} is the number of trigger hadrons in the given transverse momentum bin, here $p_{\text{T,trig}} \in [20, 50]$ GeV/c and $[8, 9]$ GeV/c, respectively. By construction, Δ_{recoil} does not contain the yield of background jets. For both trigger hadron p_{T} bins, this contribution is the same and is removed by the subtraction. Let us also add that the Δ_{recoil} observable does not impose any bias on the jet fragmentation.

The Δ_{recoil} distribution, fully corrected for the effects related to detector response and background fluctuations, can be used to study the medium induced modification of jet fragmentation. One possibility is to compare the yield of recoil jets in Pb–Pb and pp collisions using the ratio

$$\Delta I_{AA} = \Delta_{\text{recoil}}^{\text{Pb-Pb}} / \Delta_{\text{recoil}}^{\text{pp}} \quad (2)$$

where $\Delta_{\text{recoil}}^{\text{Pb-Pb}}$ spectrum obtained in Pb–Pb system is divided by the reference $\Delta_{\text{recoil}}^{\text{pp}}$ data measured in pp collisions at the same center-of-mass energy per nucleon pair. Alternatively one can search for the modification of transverse jet structure based on the ratio of two $\Delta_{\text{recoil}}^{\text{Pb-Pb}}$ distributions corresponding to different R.

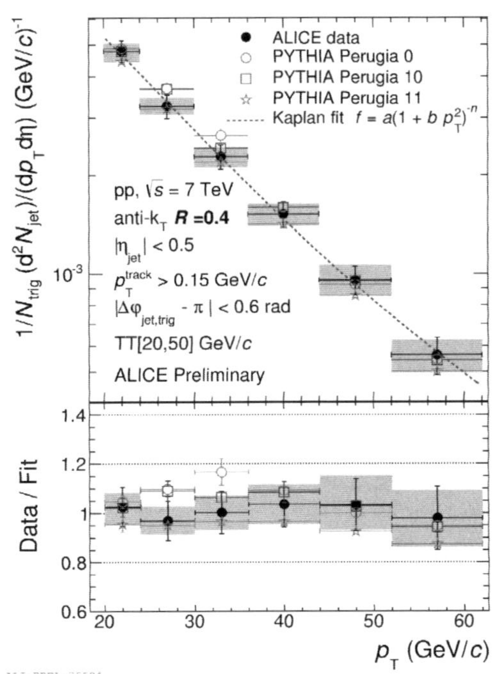

Figure 2. Top: Per trigger normalized p_{T} spectrum of anti-k$_{\text{t}}$ $R = 0.4$ recoil charged jets associated with a [20,50] GeV/c trigger hadron in pp collisions at $\sqrt{s} = 7$ TeV. Systematic uncertainties on the measured data are marked by the gray boxes. The ALICE data are compared to three PYTHIA Perugia tunes 0, 10, and 11. Bottom: Ratio of the data sets w.r.t. to the Kaplan function fit of the measured data.

Owing to the limited statistics of the pp data at $\sqrt{s} = 2.76$ TeV the reference spectrum was estimated by means of the PYTHIA event generator using Perugia 10 tune [9]. To validate this approach PYTHIA predictions were tested on the measured semi-inclusive recoil jet yield from pp at $\sqrt{s} = 7$ TeV. This data set contains 168 M minimum bias events and was taken in 2010. The analysis of recoil jet yields in pp collisions at $\sqrt{s} = 7$ TeV closely follows the Pb–Pb data analysis. Underlying event activity was estimated in a $R = 0.4$ cone rotated by 90 degrees

IOP Conf. Series: Journal of Physics: Conf. Series **805** (2017) 012013 doi:10.1088/1742-6596/805/1/012013

in azimuth w.r.t. the leading jet in event. As can be seen from Fig. 2 and Figs. 3 and 4, the PYTHIA Perugia tunes reproduce the fully corrected measured jet spectrum as well as the ratios of recoil jet yields corresponding to different R. Our studies suggest that the Perugia 10 tune provides the best description of the data.

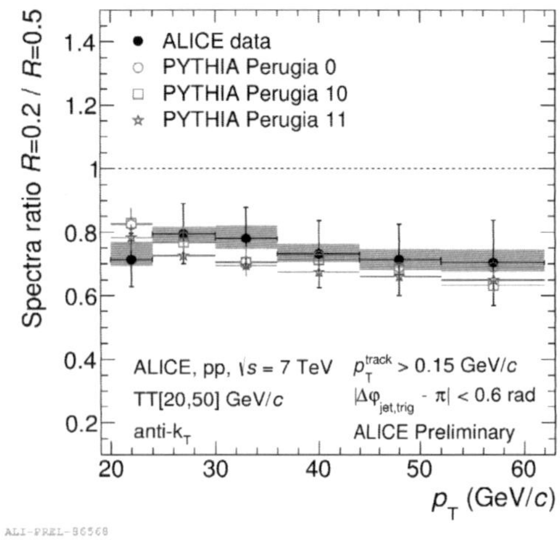

Figure 3. Ratio of per trigger normalized p_T spectra of charged anti-k_t $R = 0.2$ and $R = 0.5$ jets associated with a [20,50] GeV/c trigger hadron in pp collisions at $\sqrt{s} = 7$ TeV. Gray boxes mark the systematic uncertainties on the measured data. The ALICE data are compared to PYTHIA Perugia tunes.

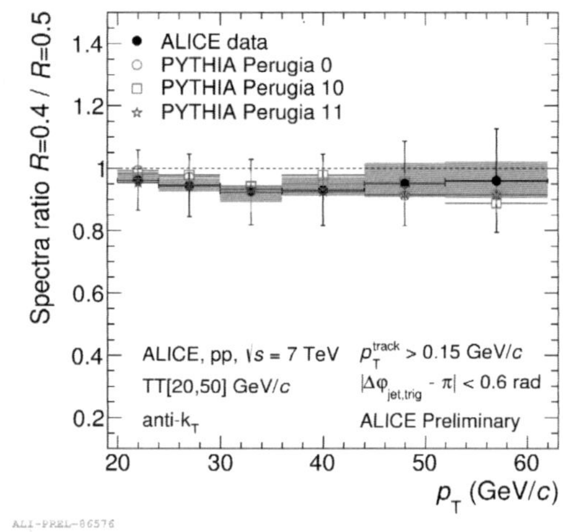

Figure 4. The same as Fig. 3 but for $R = 0.4$ and $R = 0.5$ jets.

Using a pp reference from PYTHIA Perugia 10 we evaluated the ΔI_{AA} ratio, see Fig. 5. The observed suppression of the recoil jet yield in Pb–Pb relative to pp is a consequence of the jet quenching phenomenon. Within the statistical and systematic uncertainties no p_T dependence of ΔI_{AA} is observed.

As noted above, the ratio of two $\Delta_{\rm recoil}^{\rm Pb\text{-}Pb}$ distributions that correspond to different R carries information about the transverse jet structure. The magnitude of the medium induced modification can be inferred from a comparison of the measured ratio with the data representing the vacuum fragmentation, see Fig. 6. The vacuum fragmentation is estimated by the PYTHIA Perugia tunes. The measured data do not exhibit any significant energy redistribution w.r.t. PYTHIA which means that there is no evidence for intra-jet broadening within $R = 0.5$.

To conclude, hadron-jet correlation observables represent a promising way to study the modification of jet fragmentation in medium. The main advantage of this approach is that it does not impose a fragmentation bias on the recoil jet and in addition it allows to study low p_T jets with large radii and with a minimum infrared cutoff.

The work has been supported by the MEYS grant CZ.1.07/2.3.00/20.0207 of the European Social Fund (ESF) in the Czech Republic: Education for Competitiveness Operational Programme (ECOP) and by the grant LG 13031 of the Ministry of Education of the Czech Republic.

References

[1] de Barros G O V, Fenton-Olsen B, Jacobs P, Ploskon 2013 *Nucl. Phys. A* **910–911** 314

IOP Conf. Series: Journal of Physics: Conf. Series **805** (2017) 012013 doi:10.1088/1742-6596/805/1/012013

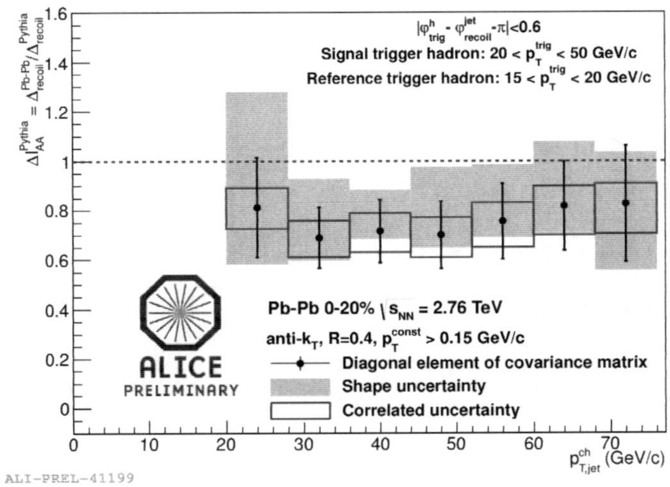

Figure 5. ΔI_{AA} corresponding to anti-k_t $R = 0.4$ jets in Pb–Pb at $\sqrt{s_{NN}} = 2.76$ TeV. The pp reference was estimated by PYTHIA Perugia 10 tune.

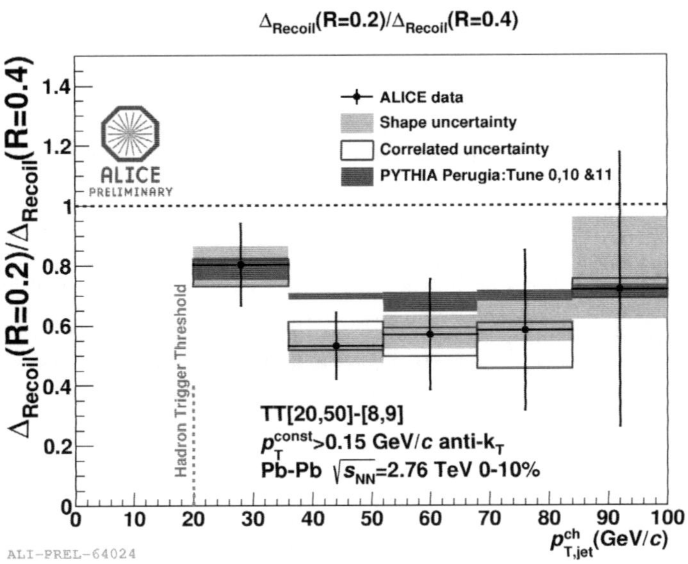

Figure 6. Ratio of two $\Delta_{\rm recoil}$ spectra corresponding to different jet R, here $R = 0.2$ and $R = 0.4$. Measured data are from 0-10% centrality bin of Pb–Pb collisions at $\sqrt{s_{NN}} = 2.76$ TeV. The red band shows the spread of predictions of this ratio by the PYTHIA Perugia tunes and represents the trend expected from vacuum fragmentation.

[2] ALICE collaboration 2008 *J. Instrum.* **3** S08002
[3] Cacciari M, Salam G P, Soyez G 2012 *Eur. Phys. J. C* **72** 1896
[4] Cacciari M and Salam G P 2008 *Phys. Lett. B* **659** 119
[5] Höcker A and Kartvelishvili V 1996 *Nucl. Instrum. Meth. A* **372** 469
[6] D'Agostini G 1995 *Nucl. Instrum. Meth. A* **362** 487
[7] Adye T *CERN–2011–006* 313
[8] Renk T 2013 *Phys. Rev. C* **87** 2
[9] Skands P 2010 *Phys. Rev. D* **82** 074018

IOP Conf. Series: Journal of Physics: Conf. Series **805** (2017) 012014 doi:10.1088/1742-6596/805/1/012014

Identified Two-particle Correlations and Quantum Number Conservations in p-p and Pb-Pb Collisions at LHC Energies

Gyula Bencédi[1], **Gergely Gábor Barnaföldi**[1], **Levente Molnár**[2]

[1] Wigner Research Centre for Physics of the Hungarian Academy of Sciences, Budapest, Hungary

[2] Institut Pluridisciplinaire Hubert Curien, Strasbourg, France

E-mail: `bencedi.gyula@wigner.mta.hu`

Abstract. In this paper we continue the investigation of the effect of quantum number conservations of pions, kaons, and protons, with very high transverse momenta (up to 25 GeV/c), during parton fragmentation and hadronization in p-p and Pb-Pb collisions at LHC energies [1, 2]. The strength of the conservation effects are studied by identified two-particle correlations in Monte Carlo generated events in the mid-rapidity region ($|\eta| < 1$). The simulated p-p events were generated with PYTHIA 8 [5], using its main default settings, at $\sqrt{s} = 200$ GeV, $\sqrt{s} = 2.76$ TeV, $\sqrt{s} = 7$ TeV, and $\sqrt{s} = 14$ TeV. In parallel to this, HIJING 1.36 [6] was used to generate Pb-Pb events at $\sqrt{s_{\rm NN}} = 2.76$ TeV with centralities $0-10\%$, $30-40\%$ and $80-90\%$. The extracted identified associated hadron spectra for charged pion, kaon, and proton show identified trigger-hadron dependent splitting between oppositely charged associated particle species in any nucleus-nucleus collisions. The Pb-Pb data exhibits a peculiar splitting pattern as a function of the transverse momentum of the associated particle $p_{T,assoc}$ both on the near and away side that is different from the patterns observed in p-p collisions. The splitting shows smooth evolution with collision energy and event multiplicity in p-p collisions while in Pb-Pb collisions different trend were observed for kaons and protons.

1. Introduction

Two-particle angular correlation is an efficient and powerful tool to study the properties of the underlying physics processes of particle production in different collision systems, like proton-proton or heavy-ion collisions. Using identified particles we can check those kind of conservation laws which are related to the different quark and flavour content of the processes [2]. This can help us to understand the specific hadronization processes in more details. Furthermore, focusing on the correlations of identified particles one can study recombination effects [3, 4] besides the parton fragmentation in the hadronization mechanisms. We used in our analysis the most known Monte Carlo generators (PYTHIA [5] and HIJING [6]) of the high energy physics community. This can be practically helpful to verify the validity and the prediction power of the studied physical quantities of this study to experimental data.

In this paper we continue our earlier work [2] on the investigation of the particle production and the quantum number conservation of identified hadron-triggered angular correlation of identified hadrons at high transverse momenta, up to $p_T = 25$ GeV/c. This time the strength of the conservation effects is studied as a function of collision energy and event multiplicity in p-p collisions and as a function of centrality in Pb-Pb collisions.

This proceeding is organized as follows: Sec. 2 briefly presents the Monte Carlo data samples and analysis settings which were used in this study. In Sec. 3 we show the obtained results for relative yields of unlike-sign associated particle pairs in specific physical configurations in p-p and Pb-Pb collisions. Finally, Sec. 4 summarizes the results of this proceeding and gives outlook for possible further analysis of the subject.

2. Monte Carlo simulations of identified two-particle azimuthal correlations in p-p and Pb-Pb collisions

2.1. Monte Carlo data sets and analysis cuts

As in our earlier study, here we also consider generated identified particles: $\pi^{\pm}$, $K^{\pm}$, p, $\bar{p}$, and the charged hadrons ($h^{\pm}$) as a reference [2].

Event generators PYTHIA 8 [5] and HIJING 1.36 [6] were used for the production of p-p and Pb-Pb event samples. PYTHIA 8 was used with its default settings (tune 4C), while HIJING event generation included the simulated effects of quenching and shadowing. The detailed data sets for the generation of p-p and Pb-Pb samples can be found in Table 1. All charged final-state particles were kept for analysis from the Monte Carlo events, nevertheless the trigger particles and associated particles used for the two-particle correlations measurements were limited in pseudorapidity in order to take into account the sensitive geometry of today's experimental detectors. Trigger particles with $|\eta_{trig}| > 0.5$ and the associated particles with $|\eta_{assoc}| > 1$ were rejected. In order to exclude the low-momentum region from the bulk effects, the transverse momenta of the trigger particles were considered above $p_{T,trig} > 2$ GeV/c in all the cases. The associated particle momenta were chosen such that $p_{T,assoc} < p_{T,trig}$ to avoid double counting. The uncorrelated background was subtracted by applying the ZYAM method [7].

Collision system	$\sqrt{s_{NN}}$	Statistics
p-p	14 TeV	500 M
p-p	7 TeV	100 M
p-p	2.76 TeV	100 M
p-p	200 GeV	500 M
Pb-Pb (0-10%)	2.76 TeV	4 M
Pb-Pb (30-40%)	2.76 TeV	2 M
Pb-Pb (80-90%)	2.76 TeV	10 M

Table 1. The list of Monte Carlo generated data sets used for our analysis. Proton-proton events were generated with different collision energies, while Pb-Pb events simulated with different centralities as indicated in the parenthesis. See the text for further details.

2.2. The definition of the associated per trigger yield

Measuring quantum number conservation requires constraints both on the trigger and the associated particle directions to tighten the sensitivity of the measurement. The associated per trigger yield is defined with the following quantity as a function of pseudorapidity difference $\Delta\eta = \eta_{trig} - \eta_{assoc}$ and azimuthal angle difference $\Delta\phi = \phi_{trig} - \phi_{assoc}$:

$$\frac{\mathrm{d}^2 N}{\mathrm{d}(\Delta\eta)\mathrm{d}(\Delta\phi)}(\Delta\phi, \Delta\eta, p_{T,trig}, p_{T,assoc}) = \frac{1}{N_{\mathrm{trig}}} \cdot \frac{\mathrm{d}N_{\mathrm{assoc}}}{\mathrm{d}(\Delta\eta)\mathrm{d}(\Delta\phi)}, \quad (1)$$

where N_{trig} and N_{assoc} is the number of trigger and associated particles per event, respectively. This quantity was measured in several $p_{T,trig}$ and $p_{T,assoc}$ intervals. The azimuthal angle correlations in this study are always projected within the pseudorapidity difference $|\Delta\eta| < 1$. The associated per trigger yield on the near side was extracted from the $\Delta\phi$ interval $|\Delta\phi| < \pi/2$ and the away side from the $\Delta\phi$ interval $\pi/2 < \Delta\phi < 3\pi/2$.

2.3. The PID-associated spectra

To test the quantum number conservation the identified associated spectra of the associated particles (hereafter called PID-associated spectra) have been plotted. We have investigated the PID-associated raw spectra using $\pi^{\pm}$, $K^{\pm}$, p, $\bar{p}$, and charged hadron $h^{\pm}$ triggers.

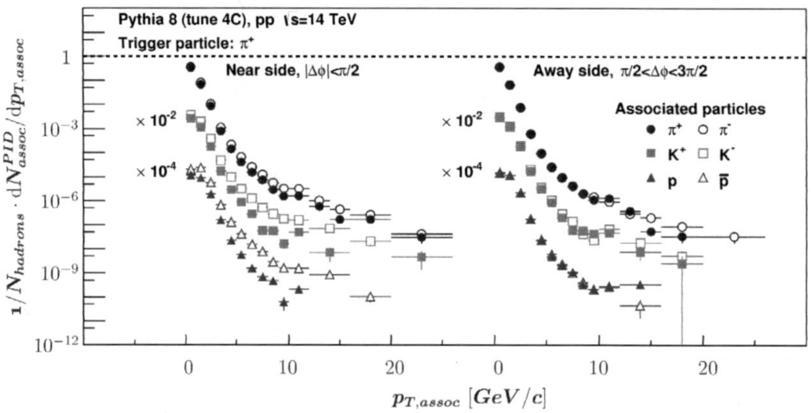

Figure 1. (Color online) π^+-triggered associated particle spectra for near- and away side particles scaled by integral of the charged hadron identified associated spectra in simulated p-p events at $\sqrt{s} = 14$ TeV. Additional scaling was applied to the spectra for better visibility. See the text for the details.

Figure 1 shows the identified particle yields $dN^{PID}_{assoc}/dp_{T,assoc}$ for the identified associated particles up to $p_{T,assoc} < 26$ GeV/c in p-p collisions at $\sqrt{s} = 14$ TeV. The kaon (green symbols) and proton (red symbols) spectra are scaled for better visibility. The PID-associated spectra have been plotted for positively charged pions, as trigger particles, selected from the transverse momentum range 2 GeV/c $< p_{T,trig} <$ 25 GeV/c.

On the left side of Figure 1 one can see the $p_{T,assoc}$ spectra for pions, kaons and protons for the 'near side' case. Right side shows the spectra for the 'away side' case. Full symbols refer to particles with positive charge and negatively charged particles are indicated with open symbols.

The PID-associated spectra of the different associated particle species show the expected evolution with the transverse momentum. The yields of the PID-associated spectra significantly decrease with the selection of charged pion, kaon and proton triggers.

On a qualitative basis one can see on the left side of Fig. 1 the yield differences between the positive and negative associated particles. Looking at the right side of the figure, no clear difference can be observed between associated particles on the away side. We listed these qualitative observations in Table 2. It shows essentially that the largest yield difference we observe is for protons and the smallest one is present for pions, as open and full symbols are splitting towards higher p_T values.

We note here, that similar plots can also be obtained using other type of trigger hadron species, as we do in the next subsection 2.4.

Trigger / Assoc	Strength of effect	
	Near side	Away side
π^+ / (π^+, π^-)	~ 2	~ 1
K^+ / (K^+, K^-)	~ 5	~ 1
p / $(p, \bar{p})$	~ 10	~ 1

Table 2. Qualitative yield differences between positive and negative associated particles on the near- and away sides, listed here on Fig. 1.

2.4. Identified particle ratios

This part was already discussed in Ref. [2] in detail, thus here we remind the reader of the main definition and features of the defined quantity.

The charge (C) and flavour specific or other quantum numbers, like baryon number (B) and strangeness (S), of generated particles are expected to hold during the hadronization and its traces can be seen on the final PID-associated spectra. In order to quantify and magnify the expected quantum number conservation effects, we define the following ratio:

$$\frac{\mathrm{d}N_{assoc}^{PID}}{\mathrm{d}p_{T,assoc}} \Big/ \frac{\mathrm{d}N_{assoc}^{hadron}}{\mathrm{d}p_{T,assoc}} \,. \tag{2}$$

This quantity can be understood as follows: in the nominator we consider the case when we are interested in the yield of the identified associated particles having an identified particle as trigger. While in the denominator we calculate the yield of identified associated particles without any PID information on the trigger particles, i.e. using charged hadrons as triggers.

In Figure 2 the ratios, defined by Eq. (2), for six identified trigger / identified associated particle species are shown: π^+/π^+, π^+/π^-, K^+/K^+, K^+/K^-, p/p, and $p/\bar{p}$. We used only the like-sign $(+/-)$, unlike-sign $(+/+)$ particle combinations, since we are interested in the flavour balance between the associated particles regarding specific quantum number.

The upper panel shows the ratio of the yields defined by Eq. (2) in p-p collisions for the near- and away side, on the left and right respectively. An obvious splitting effect can be visible between the unlike-sign associated particle pairs, which is the largest for $p/\bar{p}$, and the smallest for π^+/π^-. In contrast, on the away side the splitting effect is not really present. This indicates that in PYTHIA generated particles does not have connection to the choosen trigger particle in this context. From this picture we can qulitativley say that the strength of the quantum number conservation effects is increasing in the order of charge (C), strangeness (S) and baryon number (B) respectively.

Eq. (2) is plotted for HIJING generated events in the lower panels of Fig. 2, as well. A peculiar pattern is present on the near- and away side in the $p_{T,assoc} = 2-8$ GeV/c and $p_{T,assoc} = 2-10$ GeV/c ranges for charged kaons and protons, where the baryon/meson anomaly was observed at RHIC and at the LHC [8]. An opposite trend of the splitting at higher transverse momenta can be attributed to the minijet production embedded in the generator.

2.5. Identified particle ratios in different trigger p_T bins

So far the associated particle spectra were studied in case of wide range of trigger particles' momenta 2 GeV/c $< p_{T,trig} <$ 25 GeV/c. Here we briefly review the $p_{T,trig}$ dependence of the spectra. For this we plot the differences between the like-sign and unlike-sign particle pairs for the normalized yields $(\mathrm{d}N_{assoc}^{PID}/\mathrm{d}p_{T,assoc})/(\mathrm{d}N_{assoc}^{hadron}/\mathrm{d}p_{T,assoc})$ as plotted in Fig. 2. Since positively charged triggers were choosen we can introduce the definition below in order to plot these ratios higher than one:

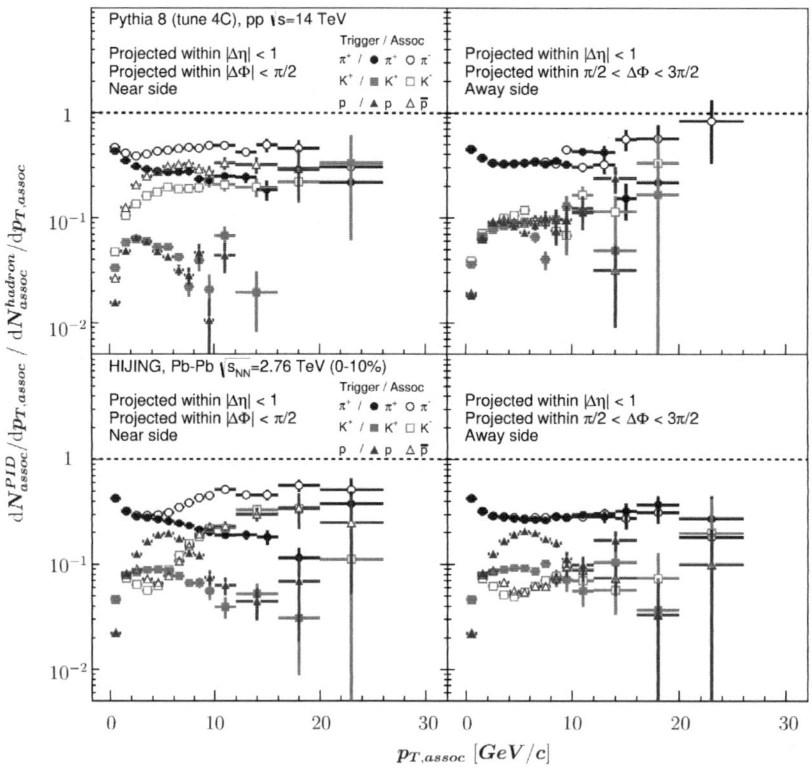

Figure 2. (Color online) PID-triggered associated particle yields relative to the charged hadron-triggered associated yields. The upper row shows the PYTHIA simulation in p-p and the lower row shows the $0 - 10\%$ central Pb-Pb HIJING events. The *left* column corresponds to the near side ($\pi/2 < |\Delta\phi|$) correlations and the *right* column corresponds to the away side ($\pi/2 < \Delta\phi < 3\pi/2$) correlations. See the text for details.

$$\left(\frac{\mathrm{d}N^{PID}_{assoc}}{\mathrm{d}p_{T,assoc}} \Big/ \frac{\mathrm{d}N^{hadron}_{assoc}}{\mathrm{d}p_{T,assoc}}\right)^{(+/-)} \Big/ \left(\frac{\mathrm{d}N^{PID}_{assoc}}{\mathrm{d}p_{T,assoc}} \Big/ \frac{\mathrm{d}N^{hadron}_{assoc}}{\mathrm{d}p_{T,assoc}}\right)^{(+/+)} = \left(\frac{\mathrm{d}N^{PID}_{assoc}}{\mathrm{d}p_{T,assoc}}\right)^{(+/-)} \Big/ \left(\frac{\mathrm{d}N^{PID}_{assoc}}{\mathrm{d}p_{T,assoc}}\right)^{(+/+)}, \tag{3}$$

where $(+/-)$ and $(+/+)$ denote the unlike-sign and like-sign trigger/associated particle pairs, respectively. This double ratio is plotted for four different $p_{T,trig}$ trigger bins on the horizontal axis of Fig. 3 in cases of p-p and most central (0-10%) Pb-Pb collisions. Deviation of the value from 1 means the relative difference between triggered like- and unlike-sign hadron yields.

In the upper row of Fig. 3 near side ratios are plotted. One can observe how the splits – in agreement with Fig. 2 – evolve as a function of $p_{T,assoc}$ in the near side cone. In general, the splitting effect is present in all choosen $p_{T,trig}$ bins, $3-7$ GeV/c, $7-10$ GeV/c, and $10-20$ GeV/c, with the same order of magnitude, for the specific partilce species. We can summerize the observed effects as follows:

- In p-p collisions the effect is present for pions, kaons, and protons in the near side cone and the largest for protons with a factor of 10 effect.

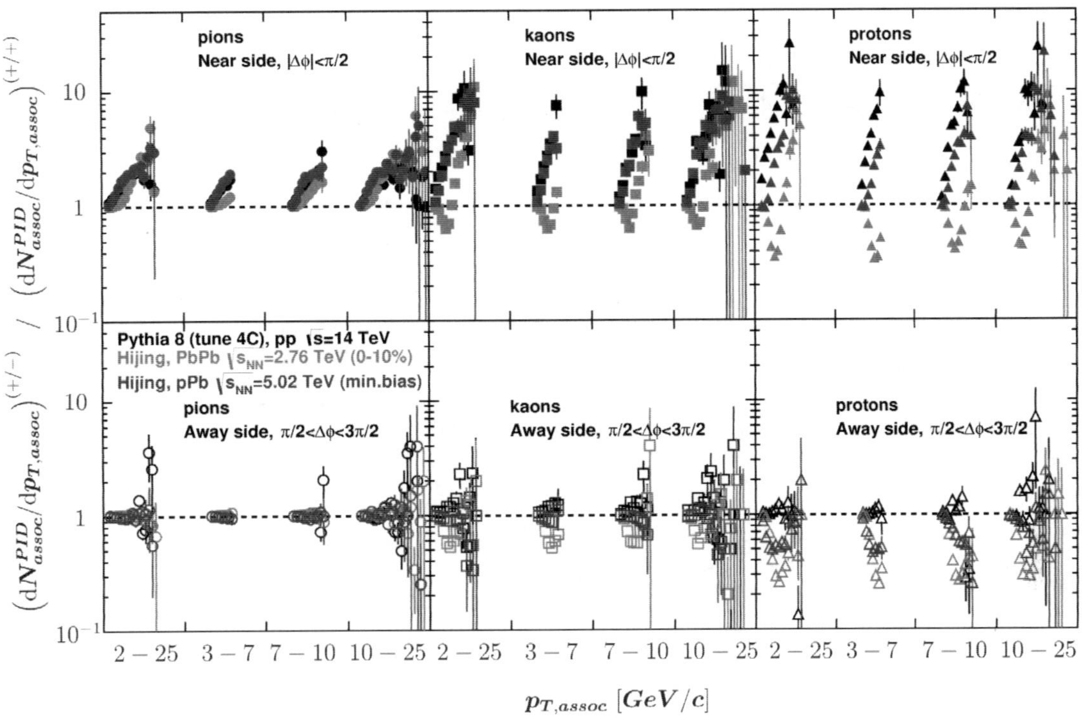

Figure 3. (Color online) Yield differences between the oppositely charged trigger/associated particle pairs. The yield enhancement is relative to the positively charged trigger particles. Different colors refer to different Monte Carlo data sets as indicated on the plot. Upper panel refers to near side, while lower panel refers to away side.

- In Pb-Pb collisions the effect is present for kaons and protons in the near- and away side cones and the largest for protons.
- In Pb-Pb collisions the observed trend of the effect for kaons and protons is completely opposite compared to the situatuion in p-p collisions in the intermediate associated transverse momentum spectra 3 GeV/c $< p_{T,assoc} <$ 7 GeV/c interval.

In p-p collisions at the away side the effect is increasing with p_T but the magnitude saturates to a constant value. A similar, but larger effect is seen in central Pb-Pb collisions within the $3 - 10$ GeV/c transverse momentum range, where the Cronin effect [9] and the baryon/meson anomaly act.

3. PID-triggered unlike sign-to-like sign associated spectra ratios in p-p and Pb-Pb collisions

3.1. Energy dependence in p-p collisions

In Fig. 4 we plotted the unlike-to-like sign relative yield ratio defined by Eq. (3) for pions, kaons, and protons with different collision energies $\sqrt{s}$. At the near side (upper panel of the figure) one can see a clear dependence of the splitting on $\sqrt{s}$ as a function of $p_{T,assoc}$ for all particle species. As we go up with $\sqrt{s}$ the effect shows a decreasing trend. This can be related to the fact that the higher the collision energy the more the number of hard interactions. This has effect on the splitting and it is 'washed out'.

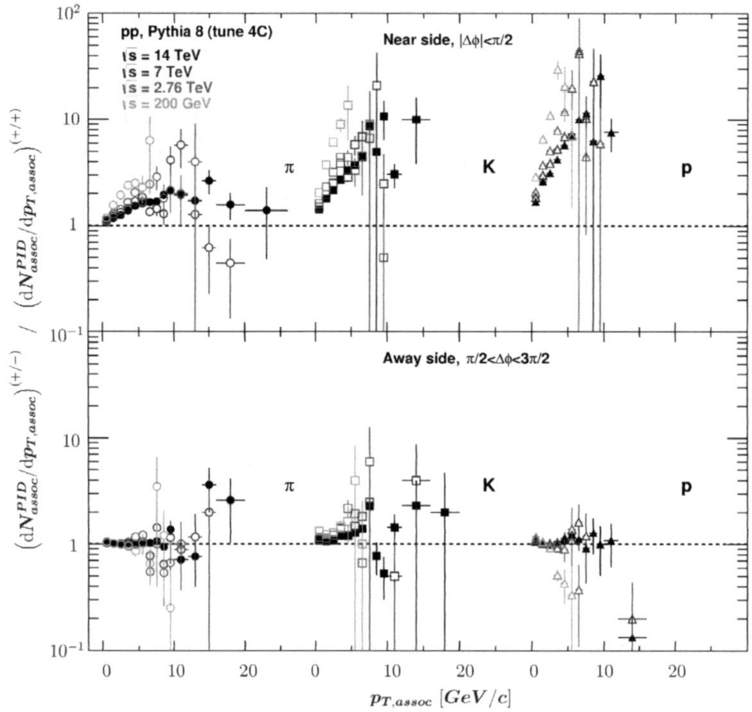

Figure 4. (Color online) Yield differences between the oppositely charged trigger/associated particle pairs for different collision energies.

On the away side we have no such kind of $\sqrt{s}$ dependence, which can be expected based on the argument we had in Sec. 2.

IOP Conf. Series: Journal of Physics: Conf. Series **805** (2017) 012014 doi:10.1088/1742-6596/805/1/012014

3.2. Event multiplicity dependence in p-p collisions

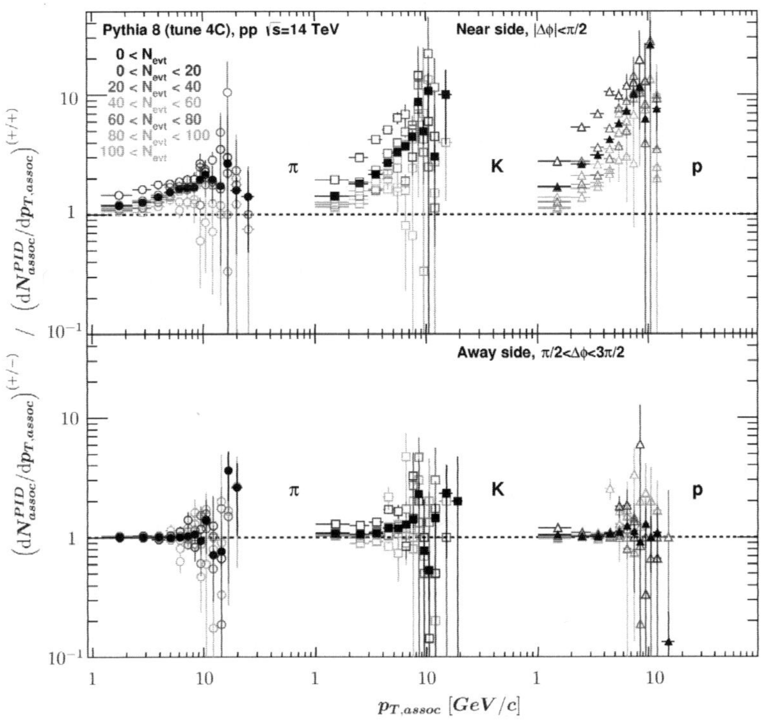

Figure 5. (Color online) Yield differences between the oppositely charged trigger/associated particle pairs for different event multiplicities in $\sqrt{s}$ = 14 TeV p-p collisions simulated by PYTHIA 8 (tune 4C).

In Fig. 5 one can see the event multiplicity dependence of Eq. (3) for pions, kaons, and protons at $\sqrt{s} = 14$ TeV in p-p collisions. We observe the same kind of trend in the evolution of the splitting with the multiplicity what have been seen in the previous case, since the number of generated particles depends on the collision energy $\sqrt{s}$. This case can be treated as a subset of the previous one, becasue here we fix the $\sqrt{s}$ and this way the generated number of particles are bouded. We can conclude that the higher the event multiplicity the smaller the size of the splitting due to increasing underlying event contribution at higher multiplicities.
Very similar to the case with $\sqrt{s}$ dependence we state here that the effects on the away side have no obvious dependence with event multiplicity.

3.3. Centrality dependence in Pb-Pb collisions

In Fig. 6 the centrality dependence of Eq. (3) is plotted in Pb-Pb collisions at $\sqrt{s_{NN}} = 2.76$ TeV. We expected here a similar effect such as obtained in the previous subsection due to the fact that the multiplicity is related to centrality. In peripheral collisions (black symbols, centrality $80 - 90\%$) we observe a very similar behaviour for pions as we have seen in Subsec. 3.1 for the collision energy dependence in p-p collisions. For kaons and protons there are two major deviations. Firstly, Eq. (3) on the near side both for the kaons and protons starts at the same value as for pions, while in Fig. 4 at the same collision energy ($\sqrt{s} = 2.76$ TeV) we see an obvious inverse deviation from unity in the first $p_{T,assoc}$ bin – which has the highest value for protons. This deviation turns to be greater than 1, approaching to highest p_T values. Secondly, we do

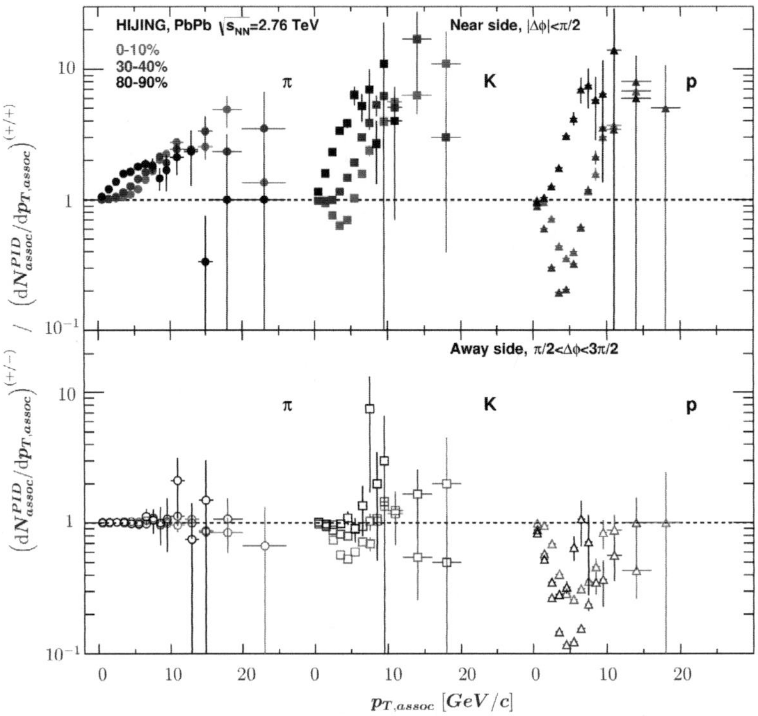

Figure 6. (Color online) Yield differences between the oppositely charged trigger/associated particle pairs for different centralities in Pb-Pb collisions at $\sqrt{s_{NN}} = 2.76$ TeV.

see an inverse evolution too on the away side for kaons and protons.
Examining the most central (red symbols, centrality $0-10\%$) and mid central (blue symbols, centrality $30-40\%$) events we clearly see a reverse trend up to $7-10$ GeV/c, which is attributed to the region of the minijet production. Surprisingly, associated particles on the away side show a stronger effect.

4. Summary and Outlook

Monte Carlo studies of two-particle azimuthal correlations have been performed with identified trigger particles and identified associated particles in p-p and Pb-Pb collisions at mid-rapidity.

The presented analysis aims to suggest a possible measurement of the charge conservation effect via PID-triggered spectra. We think it would be interesting to see whether measured data behaves similarly as PYTHIA or HIJING based Monte Carlo generated events. The key issues we have found in our analysis:

- Quantum number conservation can be seen in the near side and we have got weak signal on the away side, which suggest the correlations mainly arise from parton cascade and hadronization processes in Monte Carlo generated data samples.
- The effect is getting stronger with multiple quantum number conservation ~ 2 for pions (C), ~ 5 for kaons $(C+S)$, and ~ 10 for proton $(C+B)$.
- The relative PID-triggered yields defined by Eq. (3), presents increasing splitting up to $p_T < 10$ GeV/c and seem saturating at the highest momenta at the near side, however the lack of this effect is observed at the away side in PYTHIA simulated p-p collisions.

- In case of HIJING-simulated Pb-Pb collisions this splitting effect is also presents, but inverse trend can be seen in the unlike-to-like relative PID-triggered yield. This anomalous behavior can be either due to the minijet production or high-energy nuclear effect (Cronin, shadowing/anti-shadowing) modeled in the HIJING Monte Carlo generator.
- Both collision energy and multiplicity dependence of the effect were investigated in p-p in addition to the centrality-dependence simulations in central and peripheral Pb-Pb collisions. Here high-momenta or low-luminosity events presented larger effect, but this may sensitive to the background subtraction (here ZYAM) or the mixture of the soft-hard regime in Monte Carlo event generator cocktails.

In summary our aim is to suggest to perform the presented analysis on real data, where either the nature of the correlations or the goodness of the Monte Carlo generators can be tested.

Acknowledgement

This work was supported by Hungarian OTKA grants, NK106119, K104260, and TET 12 CN-1-2012-0016. Author Gergely Gábor Barnaföldi also thanks the János Bolyai Research Scholarship of the Hungarian Academy of Sciences.

References

[1] **ALICE** Collaboration, T. Acconcia *et al.*, "A Very High Momentum Particle Identification Detector," `http://arxiv.org/abs/1309.5880arXiv:1309.5880 [nucl-ex]`.
[2] G. Bencédi, G. G. Barnaföldi, and L. Molnár, "Monte Carlo Studies of Identified Two-particle Correlations in p-p and Pb-Pb Collisions," `http://arxiv.org/abs/1403.0117arXiv:1403.0117 [hep-ex]`.
[3] R. J. Fries, B. Muller, C. Nonaka and S. A. Bass, Phys. Rev. C **68**, 044902 (2003) [nucl-th/0306027].
[4] R. C. Hwa and C. B. Yang, Phys. Rev. C **70**, 024904 (2004) [hep-ph/0312271].
[5] T. Sjostrand, S. Mrenna and P. Z. Skands, Comput. Phys. Commun. **178**, 852 (2008) [arXiv:0710.3820 [hep-ph]].
[6] X.-N. Wang, M. Gyulassy, Phys. Rev. **D44** (1991), 3501 – 3516.
[7] A. Adare *et al.* [PHENIX Collaboration], Phys. Rev. C **78** (2008) 014901 [arXiv:0801.4545 [nucl-ex]];
[8] H. Appelshauser, J. Phys. G **38**, 124014 (2011)
[9] Antreasyan, D. *et al.* Phys. Rev. Lett. **38** (1977)
[10] S. Chatrchyan *et al.* [CMS Collaboration], [arXiv:1310.8651 [nucl-ex]].
[11] T. Sjostrand *et al.* PYTHIA 6.4 Physics and Manual, JHEP 05 (2006) 026. arXiv:hep-ph/0603175
[12] *A Very High Momentum Particle Identification Detector (VHMPID) for ALICE, Letter of Intent*, arXiv:1309.5880 [nucl-ex]
[13] V. V. Uzhinsky, hep-ph/0312089.
[14] P. Z. Skands, Phys. Rev. **D82** (2010) 074018 arXiv:hep-ph/1005.3457

Institute of Physics
Dirac House, Temple Back
Bristol BS1 6BE UK

ISSN: 1742-6588
ISBN 978-1-5108-3755-3

9 781510 837553

CURRAN ASSOCIATES INC.
proceedings
.com